AF706312

Reverse Osmosis

Principles, Design, and Operation

For Water Treatment Professionals

Industrial Water Treatment: Chemical Blending Science, Applications, and Safety
Copyright ©2025 by Venkatesha Perumal

All rights reserved. No part of this publication may be reproduced, distributed, or transmitted in any form or by any means, including photocopying, recording, or other electronic or mechanical methods, without the prior written permission of the publisher, except in the case of brief quotations embodied in critical reviews and certain other noncommercial uses permitted by copyright law.

Disclaimer:

This book is intended for informational purposes only and is based on the author's experience and research. While every effort has been made to ensure the accuracy and completeness of the information provided, the author and publisher make no representations or warranties of any kind, express or implied, about the completeness, accuracy, reliability, suitability, or availability with respect to the information, products, services, or related graphics contained in this book for any purpose. Any reliance you place on such information is therefore strictly at your own risk.

The information contained herein is not intended to be a substitute for professional advice, site-specific analysis, laboratory testing, or adherence to applicable safety procedures, regulations, and standards. Industrial water treatment involves potentially hazardous chemicals and processes. Always consult with qualified professionals, refer to Safety Data Sheets (SDS), follow established safety protocols, and comply with all local, national, and international regulations. The author and publisher shall not be liable for any loss, injury, or damage arising from the use of the information contained in this book.

ABOUT THE AUTHOR

Venkatesha Perumal

Chemical Engineer

SUMMARY OF EXPERTISE

Highly experienced and results-driven Chemical Expert with over 10 years dedicated to advancing the science and application of industrial water treatment chemistry. Possesses deep technical knowledge in chemical engineering, and troubleshooting across a wide spectrum of water systems and demanding industrial sectors, including power generation, manufacturing, and oil & gas. Combines rigorous analytical skills with extensive field experience to diagnose complex water-related problems and develop effective, safe, and compliant chemical solutions. Proven ability to translate complex technical concepts into practical applications and documentation.

AREAS OF DISTINCTION

- Water Treatment Chemical
- Application Engineering & Dosage Optimization
- Comprehensive System Troubleshooting & Problem Diagnosis
- Laboratory Analysis & Quality Control (ISO Standards)
- Chemical Safety, Handling & Storage

PROFESSIONAL FOCUS & EXPERIENCE

- Deep involvement in the **development and optimization of innovative chemical formulations** for scale inhibition, corrosion control, biofouling prevention, and solids removal, tailored for diverse water chemistries and operating conditions.
- **Extensive hands-on experience in the operation, maintenance, and performance optimization** of a wide range of industrial water treatment systems, including:
    - Boiler Water Systems (Low to High Pressure)
    - Cooling Water Systems (Open & Closed Loop)
    - Reverse Osmosis (RO) and Membrane Filtration Systems
    - Municipal Drinking Water Treatment Plants
    - Industrial and Municipal Wastewater Treatment Facilities
    - High-Purity Process Water Systems
- Proven capability in **diagnosing root causes of scale, corrosion, and fouling** through comprehensive on-site evaluation, system monitoring, and detailed water and deposit analysis.
- Strong commitment to **chemical safety**, including safe handling, storage, compatibility, and emergency response planning in blending and application environments.

- Skilled in **developing technical documentation**, including treatment proposals, service reports, and product data sheets, to support field operations and client communication.

TECHNICAL KNOWLEDGE AREAS

- **Water Treatment Systems:** Boilers, Cooling Towers, RO/NF/UF, DM Plants, Clarifiers, Wastewater Treatment (Primary, Secondary, Tertiary), High-Purity Water Systems.
- **Chemical Knowledge:** Scale Inhibitors, Corrosion Inhibitors, Oxidizing & Non-Oxidizing Biocides, Coagulants, Flocculants, pH Adjusters, Buffers, Oxygen Scavengers, Cleaning Chemicals, Membrane Chemicals, Specialty Additives, Industrial Solvents, Amines, Glycols, Hydrocarbons.
- **System Operations:** Troubleshooting, Performance Monitoring, Parameter Control, Chemical Dosing & Control, Equipment Inspection & Maintenance.
- **Equipment Familiarity:** Blending Tanks, Mixers, Pumps, Metering Systems, Filtration Equipment, Heat Exchangers, Evaporators, Clarifiers, Deaerators, Membrane Skids, Laboratory Instruments.

Preface

Water is the lifeblood of our planet, yet the availability of fresh, clean water has become one of the most pressing challenges of the twenty-first century. As populations grow and industrial demands intensify, the gap between water supply and demand continues to widen. In this landscape, Reverse Osmosis (RO) has emerged not just as a specialized filtration method, but as a transformative global technology—a literal "magic trick" that allows us to turn seawater into life-sustaining drinkable water and industrial wastewater into a reusable resource.

This book, **"Reverse Osmosis: Principles, Design, and Operation,"** was written with a specific goal in mind: to bridge the gap between complex membrane chemistry and the daily practicalities of the water treatment professional. While the scientific community has produced vast amounts of research on polymer science and thermodynamics, the engineers and operators on the ground often require a more integrated approach—one that connects the "why" of osmotic pressure to the "how" of troubleshooting a fouled membrane in the field.

What This Book Covers

We begin in **Chapter 1** by exploring the "What" and "Why" of RO, providing a historical perspective on how a curious observation of nature evolved into a multi-billion dollar industry. For those new to the field, we explain the "Magic Trick" of RO in simple, intuitive terms before diving into the rigorous science.

As we progress through **Chapter 2**, we establish the fundamental principles: the balancing act of nature known as osmosis, the role of semi-permeable membranes as gatekeepers, and the quantification of osmotic pressure. The book then transitions into the practical "nuts and bolts" of the industry. Professionals will find detailed guidance on:

- **Pretreatment (Chapters 3, 8, and 9):** Addressing the critical challenges of suspended solids, turbidity, and scaling ions to protect the membrane's integrity.
- **System Operation (Chapter 10):** A step-by-step walkthrough of commissioning and initial start-up procedures.
- **Maintenance and Cleaning (Chapter 11):** Understanding the chemistry of surfactants and antiscalants to extend membrane life.
- **Troubleshooting (Chapter 12):** A diagnostic guide for identifying and solving performance issues in real-time.

Finally, we look toward the future in **Chapter 17**, examining Thin-Film Nanocomposite (TFN) membranes, biomimetic technologies, and the role of Artificial Intelligence in optimizing water recovery.

The Intended Audience

This volume is designed for water treatment professionals across a spectrum of industries—from power generation and electronics manufacturing to municipal desalination and pharmaceutical production. Whether you are a design engineer looking to optimize a new RO train or an operator seeking to improve the Silt Density Index (SDI) of your feedwater, this book serves as both a comprehensive textbook and a practical field manual.

Our hope is that this book provides you with the clarity and tools necessary to master the "magic" of Reverse Osmosis, ensuring that the water treatment systems of tomorrow are more efficient, sustainable, and reliable than those of today.

CONTENTS

Chapter 1: Introduction and History of Reverse Osmosis

1.1 Introduction to Reverse Osmosis: The "What" and "Why"

Imagine you have a container of salty water. You want fresh water, but how do you get the salt out? You could boil the water and condense the steam (distillation), but that takes a lot of energy. You could use electrically charged materials to pull the salt ions out (ion exchange), but that works best for low salt concentrations and creates chemical waste. What if you could use a special kind of filter, one so fine that water molecules could pass through, but salt ions and other dissolved impurities couldn't?

That, in essence, is Reverse Osmosis (RO). It's a pressure-driven membrane separation process that removes a wide range of dissolved species – salts, minerals, metals, organics, microorganisms – from water. Think of it as the ultimate molecular sieve for water purification.

The "Why" of RO is compelling. It addresses one of humanity's most fundamental needs: access to clean, safe water. In a world facing increasing water scarcity due to population growth, industrial demand, and climate change, RO offers a powerful solution. It can:

1. **Create Freshwater**: Turn unusable sources like seawater and brackish groundwater into potable drinking water.
2. **Purify Water for Industry**: Provide the high-purity water essential for manufacturing processes in electronics, pharmaceuticals, power generation, and food & beverage industries.
3. **Treat Wastewater**: Cleanse industrial and municipal wastewater to levels suitable for discharge or, increasingly, for reuse, closing the water loop.
4. **Concentrate Solutions**: While often used for purification, RO principles can also be used to concentrate streams, for example, in the food industry (like concentrating fruit juices, though often Nanofiltration, a related process, is used).

It's a technology versatile enough for massive desalination plants serving entire cities and compact enough for under-the-sink units providing drinking water for a single family. Its significance in modern water management cannot be overstated.

1.2 The Magic Trick: How RO Works in Simple Terms

Before we dive into "Reverse" Osmosis, let's quickly understand natural Osmosis. If you place a semi-permeable membrane (one that allows water to pass but blocks salts) between pure water and salty water, something fascinating happens. Water molecules will naturally move from the pure water side to the salty water side, trying to dilute the salt solution. This natural movement creates a pressure difference across the membrane, known as Osmotic Pressure. Think of it as nature trying to reach equilibrium.

Analogy: Imagine two rooms separated by a wall with tiny doors only small people (water molecules) can fit through. One room is empty (pure water), the other is crowded with large people (salt ions) and small people (water molecules). The small people will naturally tend to move from the empty room into the crowded room until the "crowding" is more balanced. The pressure exerted by this movement is like osmotic pressure.

Reverse Osmosis, as the name suggests, reverses this natural flow. By applying hydraulic pressure to the salty water side – pressure that is greater than the natural osmotic pressure – we force the water molecules to move in the opposite direction: from the salty side, through the membrane, to the pure water side. The membrane acts like a gatekeeper, allowing water passage while blocking the vast majority of dissolved salts and other impurities.

Description: Below figure 1.1 illustrates the fundamental difference. On the left, natural osmosis drives water towards the salt solution, creating osmotic pressure. On the right, applied pressure exceeding the osmotic pressure forces water out of the salt solution, leaving salts behind. The water that passes through the membrane is called permeate (or product water), and the concentrated stream left behind is called concentrate (or reject, or brine).

This seemingly simple reversal is the heart of the technology, enabling the purification feats we'll explore next.

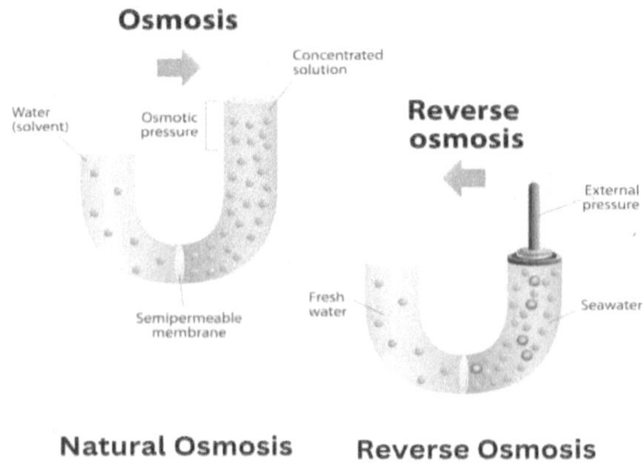

Figure 1.1: Osmosis vs. Reverse Osmosis

1.3 A World Transformed: Key Uses of Reverse Osmosis

The applications of RO are remarkably diverse, driven by its ability to remove a broad spectrum of contaminants. Here are some primary areas:

- **Desalination**: This is perhaps the most well-known use.
    - **Seawater RO (SWRO)**: Removing salts (typically 35,000-45,000 mg/L or parts per million, ppm) from ocean water to produce drinking water. This is vital for coastal communities and islands with limited freshwater resources. Early SWRO was energy-intensive, but advances in membrane technology and energy recovery devices (ERDs) have significantly reduced costs. Example: The Sorek plant in Israel, one of the world's largest and most efficient SWRO plants.
    - **Brackish Water RO (BWRO)**: Treating water with moderate salt content (typically 1,000-15,000 mg/L TDS) found in estuaries or inland aquifers. BWRO requires lower pressure than SWRO, making it less energy-intensive.

Example: Many municipalities in Florida and the US Southwest rely heavily on BWRO for their water supply.

- **Municipal Drinking Water**: Beyond desalination, RO is used to treat surface water or groundwater contaminated with specific pollutants like nitrates, arsenic, radium, or high levels of hardness or dissolved solids that other treatments struggle to remove effectively.
- **Industrial Process Water**: Many industries require water quality far exceeding typical drinking water standards.
    - Power Generation: Boiler feedwater must be exceptionally pure to prevent scale and corrosion in high-pressure boilers and turbines. RO is often a key step, frequently followed by electrodeionization (EDI) or mixed-bed ion exchange for final polishing.
    - Electronics & Semiconductor Manufacturing: Requires ultrapure water (UPW), where impurity levels are measured in parts per trillion. RO is indispensable in the multi-stage treatment train for UPW production. Even trace ionic or organic contaminants can ruin microchips.
    - Pharmaceutical Production: Water for Injection (WFI) and USP Purified Water have stringent purity requirements defined by pharmacopoeias. RO is a core technology used to meet these standards, ensuring product safety and efficacy.
    - Food and Beverage: Used for removing minerals that affect taste (e.g., in bottled water, soft drinks), concentrating products (like whey protein in the dairy industry), or providing high-quality process water for cleaning and production.
- **Wastewater Treatment and Reuse**: As water scarcity intensifies, recycling wastewater becomes crucial. RO can treat secondary or tertiary effluent from municipal or industrial wastewater plants, removing residual salts, organics, and pathogens to produce high-quality water suitable for irrigation, industrial reuse, or even indirect potable reuse (recharging aquifers). Example: Orange County Water District's Groundwater Replenishment System (GWRS) in California is a landmark project using microfiltration, RO, and UV/peroxide to purify wastewater for aquifer recharge.
- **Residential Use**: Point-of-use (POU) RO systems, typically installed under the kitchen sink, provide high-quality drinking water by removing dissolved salts, heavy metals (like lead), chlorine, and other contaminants from tap water.

- **Laboratory Use**: Labs require various grades of pure water for experiments and analyses. Benchtop RO units, often combined with other polishing technologies, are common.

Table 1.1: Major Applications of Reverse Osmosis and Typical Objectives

Application Area	Primary Objective	Typical Feed Water TDS (mg/L)	Typical Permeate TDS (mg/L)	Key Challenge(s)
Seawater Desalination	Potable Water Production	35,000 - 45,000	< 500	High Pressure, Energy, Fouling
Brackish Water Desal.	Potable/Industrial Water Production	1,000 - 15,000	< 500	Scaling, Fouling
Municipal Drinking Water	Removal of Specific Contaminants	200 - 1,500	< 50 - 500	Pretreatment, Concentrate Disposal
Power Generation	Boiler Feedwater (High Purity)	100 - 1,000	< 5 - 10	Silica Removal, High Purity
Electronics (UPW)	Ultrapure Water	50 - 500	<< 1 (ppb levels)	Extreme Purity, TOC/Particle Removal
Pharmaceuticals (PW/WFI)	USP/EP Compliance	50 - 500	< 1 - 5	Sanitization, Validation
Wastewater Reuse	Water Recovery for Reuse (Industrial, Irrigation, IPR)	500 - 5,000	< 100 - 500	High Fouling Potential (Organic/Bio)
Residential (POU)	Improved Taste, Removal of Specific Contaminants	100 - 1,000	< 50	Low Recovery, Maintenance

Note: TDS (Total Dissolved Solids) values are indicative and can vary significantly.

The breadth of these applications underscores why understanding RO is so essential in today's world.

1.4 FromCurious Observation to Global Technology: A History of RO Development

The journey of RO wasn't a single 'eureka' moment but a gradual progression built on scientific observation, theoretical understanding, and materials science breakthroughs.

1.4.1 Early Observations and Osmotic Pressure (18th - 19th Centuries): The phenomenon of osmosis was first systematically observed by Jean-Antoine Nollet in 1748 using an animal bladder as a membrane separating water and alcohol. Over the next century, scientists studied this phenomenon, leading to Jacobus Henricus van't Hoff's work in the 1880s, which established the mathematical relationship between solute concentration and osmotic pressure, earning him the first Nobel Prize in Chemistry in 1901. At this stage, osmosis was a scientific curiosity, and reversing it wasn't seriously contemplated for practical water treatment.

1.4.2 The Breakthrough: Loeb and Sourirajan (Late 1950s - Early 1960s): The modern era of RO began at the University of California, Los Angeles (UCLA). Driven by the need for desalination technologies (partly funded by the US Office of Saline Water), researchers sought materials that could act as practical semi-permeable membranes. Early attempts using materials like cellulose acetate (CA) showed some salt rejection, but the water flux (the rate of water passage) was agonizingly low – too low for any practical application.

The pivotal breakthrough came from Sidney Loeb and Srinivasa Sourirajan. In the early 1960s, they developed a method for casting cellulose acetate membranes that had a unique asymmetric structure. This membrane consisted of a very thin, dense "skin" layer (less than 0.25 micrometers thick) responsible for the salt rejection, supported by a much thicker, porous substructure that provided mechanical strength but offered little resistance to water flow.

Analogy: Think of it like Gore-Tex® fabric. A very thin waterproof-yet-breathable layer is laminated onto a thicker, stronger fabric. The thin layer does the work (blocking liquid water but allowing water vapor through), while the support layer provides durability.

This asymmetric structure dramatically increased water flux while maintaining good salt rejection, making RO a potentially viable industrial process for the first time. This Loeb-Sourirajan membrane became the foundation for the first generation of commercial RO systems.

1.4.3 The Next Leap: Thin-Film Composites (1970s): While the asymmetric CA membrane was revolutionary, it had limitations. Cellulose acetate is prone to hydrolysis (breakdown by water, especially at non-neutral pH) and biodegradation (it's essentially plant fiber), and it has limited tolerance to chlorine, a common disinfectant.

The search for better membrane materials led to the development of Thin-Film Composite (TFC) membranes, primarily pioneered by John Cadotte at the North Star Research Institute (later FilmTec Corp., now part of DuPont). TFC membranes represent a different manufacturing approach. A thin, selective polyamide layer (the "separation layer," typically only 0.1-0.2 micrometers thick) is formed in-situ via interfacial polymerization on the surface of a pre-formed microporous support layer (commonly polysulfone), which itself sits on a non-woven polyester fabric web for mechanical support.

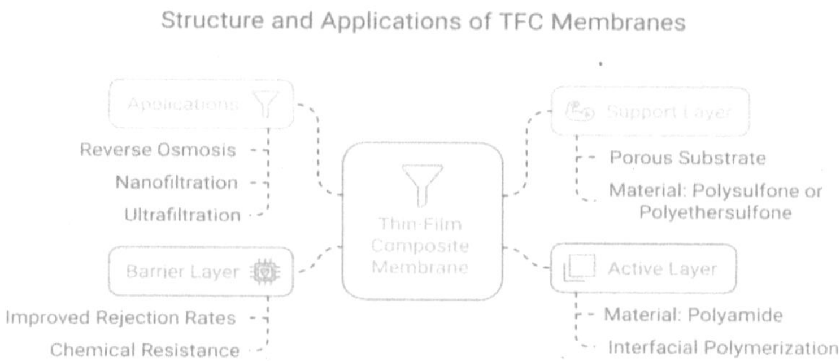

Figure 1.2: Structure and Applications of TFC Membranes

TFC membranes offered several advantages over CA membranes:

- Higher water flux at similar rejection levels.
- Significantly higher salt rejection, especially crucial for seawater desalination.
- Stability over a wider pH range (typically pH 2-11).
- Better resistance to biodegradation.

However, TFC polyamide membranes generally have very poor tolerance to strong oxidizers like chlorine, which became a major consideration for pretreatment system design (chlorine must typically be removed before the water reaches the RO membranes). Despite this limitation, TFC membranes quickly became the industry standard for most RO applications due to their superior performance, and they remain so today, albeit with many refinements.

1.4.4 Commercialization and Growth (1960s - Present): Following the Loeb-Sourirajan breakthrough, companies like Aerojet General, DuPont, Dow Chemical (through its acquisition of FilmTec), Hydranautics (originally founded based on CA technology), Toray, and others began developing and commercializing RO technology. Early applications focused on brackish water treatment. As membrane performance improved and costs decreased (especially with the advent of TFC membranes and more efficient module designs like the spiral-wound module, which we'll discuss in Chapter 4), seawater desalination became increasingly feasible. The energy crisis of the 1970s spurred interest in less energy-intensive desalination methods like RO compared to distillation. Over the past few decades, RO has seen exponential growth, becoming the dominant technology for desalination and a standard tool for high-purity water production and water reuse worldwide.

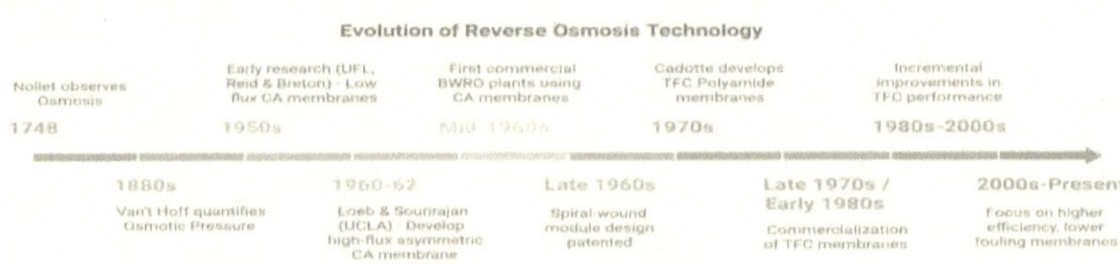

Figure 1.3: Key Milestones in Reverse Osmosis History

1.5 Constant Evolution: Recent Advances in RO Membrane Technology

While the fundamental TFC structure remains, research and development continue to yield significant improvements:

- **Higher Flux Membranes**: Allowing more water to pass through a given area of membrane at the same pressure, reducing the number of membrane elements needed and thus capital costs.
- **Higher Rejection Membranes**: Improving the removal of challenging contaminants like boron (important for irrigation water derived from SWRO) and small, uncharged organic molecules.
- **Low Energy / Low Pressure Membranes:** Designed to operate efficiently at lower feed pressures, primarily for BWRO and water reuse applications, saving significant energy.
- **Fouling Resistant Membranes:** Surface modifications (e.g., coatings, grafting hydrophilic substances, manipulating surface charge and roughness) make membranes less prone to accumulating deposits (fouling and scaling), leading to more stable operation, less frequent cleaning, and longer membrane life. This is a major area of ongoing research.
- **Chlorine Tolerant Membranes:** While TFC polyamide remains sensitive, research continues into developing membranes with better chlorine resistance to simplify pretreatment, although truly high-performing chlorine-tolerant RO membranes have proven elusive compared to standard polyamides. Some variations exist with limited tolerance.
- **Nanotechnology Integration:** Incorporating nanomaterials (e.g., zeolites, carbon nanotubes, metal-organic frameworks, graphene oxide) into the polyamide layer or support structure is being explored to enhance permeability, selectivity, fouling resistance, or mechanical strength. This is still largely in the R&D or niche application stage but holds future promise.

These advances are driven by the constant demand for lower costs (both capital and operational, especially energy), better performance, greater reliability, and the ability to tackle increasingly challenging feed waters and achieve higher levels of water recovery.

1.6 Peeking Over the Horizon: Future Advancements

The field of RO and membrane separations continues to be dynamic. Looking ahead, we can anticipate trends such as:

- **Biomimetic Membranes:** Inspired by biological cell membranes that use highly efficient protein channels called aquaporins to transport water with incredible speed and selectivity. Research aims to create synthetic membranes incorporating aquaporins or mimicking their structure for potentially ultra-high flux and high rejection.
- **Advanced Anti-Fouling Strategies:** Moving beyond passive resistance to active fouling control, perhaps using membranes with responsive surfaces or integrated sensing capabilities.
- **Improved Selectivity:** Developing membranes that can selectively remove specific target ions while allowing others to pass, potentially valuable for resource recovery from brines or tailored water conditioning.
- **Hybrid Processes:** Combining RO with other membrane processes (like Forward Osmosis - FO, Membrane Distillation - MD) or other treatment technologies to optimize energy efficiency, improve water recovery, manage difficult feed waters, or achieve specific separation goals.
- **Artificial Intelligence (AI) and Machine Learning:** Using AI for smarter process control, predictive maintenance (e.g., predicting fouling events), optimizing cleaning schedules, and potentially even aiding in the design of new membrane materials and systems.
- **Brine Management and Resource Recovery:** As RO use expands, managing the concentrated brine stream becomes increasingly important. Future developments will likely focus on technologies for brine concentration (minimizing disposal volume) and extracting valuable minerals or salts from the brine (turning waste into a resource).

1.7 Chapter Summary and Key Takeaways

- Reverse Osmosis (RO) is a pressure-driven membrane process that separates dissolved solutes from water by forcing water through a semi-permeable membrane against the natural osmotic pressure gradient.

- RO is a cornerstone technology for producing fresh water from saline sources (seawater, brackish water), generating high-purity water for various industries, and enabling wastewater reuse.
- The key historical breakthroughs were the development of the asymmetric cellulose acetate membrane by Loeb and Sourirajan, making RO practical, and the subsequent invention of the high-performance thin-film composite (TFC) polyamide membrane by Cadotte, which became the industry standard.
- Modern RO membranes are typically TFC structures with a thin polyamide rejection layer on a porous support. Ongoing research focuses on improving flux, rejection, fouling resistance, energy efficiency, and durability.
- Future trends point towards biomimetic designs, advanced anti-fouling, increased selectivity, hybrid systems, AI integration, and better brine management solutions.

1.8 References

1. Loeb, S., & Sourirajan, S. (1963). Sea water demineralization by means of an osmotic membrane. Advances in Chemistry Series, 38, 117–132. (The seminal paper on asymmetric CA membranes).
2. Cadotte, J. E. (1981). Interfacially synthesized reverse osmosis membrane. U.S. Patent No. 4,277,344. Washington, DC: U.S. Patent and Trademark Office. (Key patent for TFC membranes).
3. Baker, R. W. (2012). Membrane Technology and Applications (3rd ed.). John Wiley & Sons. (A comprehensive textbook covering various membrane processes).
4. Elimelech, M., & Phillip, W. A. (2011). The Future of Seawater Desalination: Energy, Technology, and the Environment. Science, 333(6043), 712–717. (Excellent overview of challenges and future directions, particularly for SWRO).
5. Greenlee, L. F., Lawler, D. F., Freeman, B. D., Marrot, B., & Moulin, P. (2009). Reverse osmosis desalination: Water sources, technology, and today's challenges. Water Research, 43(9), 2317–2348. (Comprehensive review of RO desalination).
6. Fritzmann, C., Löwenberg, J., Wintgens, T., & Melin, T. (2007). State-of-the-art of reverse osmosis desalination. Desalination, 216(1–3), 1–76. (Detailed review covering history, membranes, modules, pretreatment, and applications).

7. American Water Works Association (AWWA). (2007). Manual M46: Reverse Osmosis and Nanofiltration. AWWA. (Industry standard manual focusing on practical aspects).
8. DuPont Water Solutions. (Various Technical Manuals and Literature). e.g., FilmTec™ Membranes Technical Manual. (Manufacturer literature often provides practical insights into their specific products and general RO principle.

Chapter 2: Reverse Osmosis Principles

2.1 Osmosis: Nature's Balancing Act

Osmosis is a fundamental process in nature, crucial for everything from nutrient uptake in plant roots to maintaining the balance of fluids in our own bodies. It describes the spontaneous movement of solvent molecules (usually water, in our context) across a selectively permeable membrane from a region of high solvent potential (low solute concentration) to a region of low solvent potential (high solute concentration). In simpler terms: water moves towards the area where there's more "stuff" dissolved in it, trying to dilute it.

2.1.1 The Gatekeeper: Semi-Permeable Membranes

The key player in osmosis (and RO) is the semi-permeable membrane. The name says it all: "semi" meaning partly, and "permeable" meaning allowing things to pass through. It's a barrier that selectively allows certain molecules or ions to pass through it by diffusion, while blocking others.

In the context of water treatment, an ideal RO membrane would:

- Allow water molecules (H_2O) to pass through easily.
- Block dissolved salts (like Sodium Chloride, NaCl, which exists as Na^+ and Cl^- ions in water), minerals (Ca^{2+}, Mg^{2+}), metals, bacteria, viruses, and larger organic molecules.

No real-world membrane is perfectly selective, but modern RO membranes come remarkably close for many contaminants. The first membranes used (like animal bladders in early experiments, or the cellulose acetate membranes developed by Loeb and Sourirajan) and the advanced thin-film composite membranes used today all function based on this principle of selective passage. The exact mechanism of how water passes while salts are rejected is complex and debated (we'll touch on transport models in Chapter 4), but the result is selective permeability.

2.1.2 The Driving Force: Water Potential

Why does water move spontaneously during osmosis? Thermodynamics tells us that systems tend to move towards a state of lower energy or higher entropy (greater randomness/disorder). Water molecules, like everything else, prefer to be in a lower energy state.

Pure water has the highest "water potential" (think of it as water's chemical energy level or its tendency to move). When you dissolve solutes (like salt) in water, you lower the water's potential. Why?

1. **Dilution Effect**: The concentration of water molecules per unit volume decreases.
2. **Interaction Effect**: Water molecules interact with the solute ions/molecules (hydration), restricting their freedom to move compared to pure water.

When you have pure water and salty water separated by a semi-permeable membrane, water molecules will naturally move from the high-potential (pure water) side to the low-potential (salty water) side to try and equalize this potential difference. This movement continues until equilibrium is reached, or until the pressure on the salty side increases enough to counteract the potential difference.

2.1.3 Quantifying the Drive: Osmotic Pressure (π)

The Osmotic Pressure (π) is the minimum pressure that needs to be applied to the solution (the salty side) to prevent the inward flow of pure water across the semi-permeable membrane. It's a measure of the "tendency" of water to move into that solution. The higher the concentration of solutes in the solution, the lower its water potential, and the higher its osmotic pressure.

Analogy: Imagine a sponge soaking up water. Osmotic pressure is like the pressure the sponge exerts, drawing water in. To stop more water from entering, you'd have to press on the sponge (or the water surrounding it) with an equal counter-pressure.

For dilute solutions, osmotic pressure can be estimated using the Van't Hoff Equation, which bears a striking resemblance to the ideal gas law:

$\pi = i * C * R * T$

Where:

- **π (Pi)** is the Osmotic Pressure (typically in atmospheres (atm), pounds per square inch (psi), or Pascals (Pa)).
- **i** is the **Van't Hoff factor**. It represents the number of ions or particles the solute dissociates into when dissolved in the solvent. For non-dissociating solutes like sugar, i = 1. For NaCl (which splits into Na^+ and Cl^-), i ≈ 2. For $CaCl_2$ (Ca^{2+} + $2Cl^-$), i ≈ 3. (Note: This factor is slightly less than the theoretical integer value in real solutions due to ion pairing effects, especially at higher concentrations).
- **C** is the **molar concentration of the solute** (moles per liter, mol/L). To use this formula practically with water analyses often reported in mg/L (ppm), you first need to convert the concentration of each salt to moles/L.
    - Conversion: Molar Concentration (mol/L) = Concentration (mg/L) / (Molecular Weight (g/mol) * 1000 mg/g)
- **R** is the **Ideal Gas Constant**. Its value depends on the units used for pressure and volume. Common values include:
    - 0.08206 L·atm / (mol·K)
    - 8.314 L·kPa / (mol·K) (since 1 kPa ≈ 0.145 psi)
    - (For psi, using R tailored for this often involves pre-converting units)
- **T** is the **Absolute Temperature** in Kelvin (K = °C + 273.15).

Calculating Osmotic Pressure - An Example:

Let's estimate the osmotic pressure of a simple brackish water solution containing 1000 mg/L of Sodium Chloride (NaCl) at 25°C (298.15 K).

Molecular Weight of NaCl: Na (22.99 g/mol) + Cl (35.45 g/mol) = 58.44 g/mol

Molar Concentration (C): C = (1000 mg/L) / (58.44 g/mol * 1000 mg/g) ≈ 0.0171 mol/L

Van't Hoff Factor (i): For NaCl, theoretically i = 2. We'll use this for approximation.

Temperature (T): 25°C = 298.15 K

Ideal Gas Constant (R): Let's use 0.08206 L·atm/(mol·K) to get pressure in atm.

Calculate π:

π = (2) * (0.0171 mol/L) * (0.08206 L·atm/(mol·K)) * (298.15 K)

π ≈ 0.836 atm

Convert to psi: Since 1 atm ≈ 14.7 psi:

π ≈ 0.836 atm * 14.7 psi/atm ≈ 12.3 psi

Important Considerations:

- Mixtures: Real water contains multiple salts. The total osmotic pressure is approximately the sum of the osmotic pressures contributed by each individual solute.
- TDS Approximation: For rough estimates, a rule of thumb is often used: For every 100 mg/L of Total Dissolved Solids (TDS), the osmotic pressure is approximately 1 psi (at typical temperatures). This is a very rough guideline, more accurate for NaCl-dominated waters, but useful for quick checks. Using our example (1000 mg/L TDS), this rule gives 10 psi, which is reasonably close to the calculated 12.3 psi. For seawater (≈35,000 mg/L TDS), this rule suggests ≈350 psi, which is in the right ballpark (actual values are often 350-400 psi depending on specific composition and temperature).
- Concentration Effects: The Van't Hoff equation is most accurate for dilute solutions. At higher concentrations, like in seawater or RO concentrate streams, the actual osmotic

pressure might be slightly higher than predicted due to non-ideal solution behavior.

Table 2.1: Approximate Osmotic Pressure of Various Water Types (at 25°C)

Water Type	Typical TDS (mg/L)	Estimated Osmotic Pressure (psi)	Estimated Osmotic Pressure (bar)	Notes
High Purity Water	< 1	~ 0	~ 0	Negligible osmotic pressure
Typical Tap Water	200 - 500	2 - 5	0.14 - 0.34	Low
Brackish Water (Low)	1,500	~ 15	~ 1.0	Requires moderate pressure for RO
Brackish Water (High)	5,000	~ 50	~ 3.4	Significant pressure needed for RO
Seawater	35,000	~ 350 - 400	~ 24 - 27	Requires high pressure for RO
RO Concentrate (SWRO)	60,000 - 70,000	~ 600 - 750	~ 41 - 52	Very high osmotic pressure limits recovery
Great Salt Lake	$> 200,000$	> 2000	> 138	Extremely high, challenging for standard RO

Note: These are typical values. Actual osmotic pressure depends on specific ionic composition and temperature. 1 bar ≈ 14.5 psi.

2.1.4 Osmosis in Everyday Life

Once you understand osmosis, you see it everywhere:

- **Plant Roots:** Absorb water from the soil because the concentration of solutes inside the root cells is higher than in the soil water.
- **Pruned Fingers:** When you soak in a bath (freshwater), water enters your skin cells via osmosis, causing them to swell and wrinkle. In the salty ocean, the effect is less or even reversed.
- **Salting Food**: Salting meat or fish draws water out via osmosis, dehydrating it and helping to preserve it by making it inhospitable to bacteria.
- **Medical Applications**: Intravenous (IV) fluids must be "isotonic" – having the same osmotic pressure as blood – to avoid damaging blood cells. A solution with lower osmotic pressure (hypotonic) would cause cells to swell and burst, while one with higher pressure (hypertonic) would cause them to shrivel.

Understanding this natural tendency of water to move towards higher solute concentrations is critical because Reverse Osmosis is all about fighting this tendency.

2.2 Reverse Osmosis: Overcoming Nature's Tendency

Now we flip the coin. Instead of letting water flow naturally to dilute the salty side, we want to force water out of the salty side, leaving the salts behind.

2.2.1 Applying the Squeeze: Hydraulic Pressure

To reverse the natural osmotic flow, we need to apply external hydraulic pressure (mechanical pressure, usually from a pump) to the side of the membrane containing the feed water (the more concentrated solution).

- If Applied Pressure = Osmotic Pressure (π), the net movement of water across the membrane stops. The system is in equilibrium.
- If Applied Pressure > Osmotic Pressure (π), the natural osmotic flow is overcome, and water molecules are forced to move from the concentrated solution, through the membrane, to the dilute (permeate) side.

This is the core principle of Reverse Osmosis. The higher the applied pressure relative to the osmotic pressure, the greater the driving force for water to pass through the membrane.

2.2.2 The Real Push: Net Driving Pressure (NDP)

The actual force pushing water through the membrane isn't just the applied pressure. We need to account for the forces opposing this movement: the osmotic pressure of the feed solution and, to a lesser extent, the pressure on the permeate side (backpressure).

The **Net Driving Pressure (NDP)**, sometimes called Net Transmembrane Pressure (TMP) or Effective Feed Pressure, is calculated as:

NDP = Pf - Pp - $\Delta\pi$

Where:

- **NDP** is the Net Driving Pressure.
- **Pf** is the Hydraulic Pressure applied to the feed water.
- **Pp** is the Hydraulic Pressure on the permeate side (product water side). This is often low (close to atmospheric) in many systems, but can be significant if the permeate is being pumped or fed into a pressurized tank.
- **$\Delta\pi$ (Delta Pi)** is the Osmotic Pressure Difference across the membrane (πfeed - πpermeate). Since the permeate is ideally pure or has very low TDS, its osmotic pressure (πpermeate) is often negligible compared to the feed's osmotic pressure (πfeed). Therefore, $\Delta\pi$ is often approximated as simply the osmotic pressure of the feed solution at the membrane surface.

Crucially, as water passes through the membrane in an RO system, the remaining feed water becomes more concentrated, increasing its osmotic pressure (πfeed). Also, friction causes the hydraulic pressure (Pf) to drop as water flows along the membrane channel. This means the **NDP decreases along the length of the RO system**. This has major implications for system design, as we'll see later.

Practical Insight: Operators need to ensure the applied feed pressure is sufficiently high not just to overcome the initial osmotic pressure of the bulk feed water, but also the higher osmotic pressure of the concentrated water near the membrane surface and further down the flow path, while accounting for pressure drops.

2.2.3 Water Flow vs. Salt Flow: The Membrane's Job

Under sufficient NDP, water molecules are driven across the membrane. This flow of water is called Water Flux (Jw), usually measured in gallons per square foot per day (GFD) or liters per square meter per hour (LMH). Ideally, this flux is directly proportional to the NDP:

Jw = A * NDP

Where **A** is the **Water Permeability Coefficient,** a characteristic of the specific membrane material and structure (we'll revisit this in Chapter 4). Higher NDP means higher water flux.

Simultaneously, some solutes (salts) inevitably "leak" through the non-perfect membrane. This movement is called **Salt Flux (Js).** Unlike water flux, salt flux is primarily driven by the **concentration difference across the membrane (Cfeed - Cpermeate)**, and is much less dependent on pressure.

Js ≈ B * (Cfeed - Cpermeate)

Where **B** is the **Salt Permeability Coefficient** (again, a membrane characteristic).

This difference in driving forces is key to RO's effectiveness:

- Increasing applied pressure significantly increases water flux (Jw).
- Increasing applied pressure has relatively little effect on salt flux (Js).

Therefore, operating at higher pressures (higher NDP) generally results in both higher water production and better water quality (lower permeate salinity, because the salt flux stays relatively constant while the water flux increases, diluting the salts that do pass). However, operating at excessively high pressures increases energy costs, can physically compact some membranes, and exacerbates fouling.

2.2.4 Key Factors Influencing RO Performance

Several operational parameters interact to determine how well an RO system performs:

1. **Applied Pressure (Pf):** As discussed, higher pressure increases NDP, leading to higher water flux and generally better salt rejection (up to a point).
2. **Feed Water Concentration (TDS)**: Higher feed TDS means higher osmotic pressure (π). This reduces the NDP for a given applied pressure, thus lowering water flux. It also increases the concentration difference driving salt flux, potentially leading to slightly lower salt rejection (higher permeate TDS). Treating seawater requires much higher pressure than treating brackish water.
3. **Temperature:** Higher feed water temperature increases water flux because water viscosity decreases, making it easier for water molecules to pass through the membrane (the water permeability coefficient 'A' increases with temperature). However, higher temperature also slightly increases salt flux (the salt permeability coefficient 'B' also increases). Most manufacturers normalize performance data to 25°C (77°F). Practical Insight: Colder feed water significantly reduces permeate production rate; warmer water increases it but may slightly worsen permeate quality. Systems must be designed for the expected temperature range.

4. **Recovery Rate**: This is the percentage of feed water that is converted into permeate (product water). Higher recovery means the concentrate stream becomes much more concentrated, significantly increasing its osmotic pressure and the potential for scaling (precipitation of salts). This increased osmotic pressure reduces the NDP towards the end of the system, lowering flux in the final elements. Recovery rate is a critical design and operational parameter (discussed further in Chapter 3).

2.3 Filtration Modes: Dead-End vs. Cross-Flow

How the feed water approaches the membrane surface dramatically impacts performance, especially concerning fouling. Reverse Osmosis relies on a specific mode called cross-flow filtration.

2.3.1 Dead-End Filtration: The Coffee Filter Approach

In Dead-End Filtration (DEF), the entire fluid flow is directed perpendicularly through the filter medium. All particles or substances larger than the filter's pore size are trapped on the surface or within the filter matrix, while the filtered fluid (filtrate) passes through.

Analogy: Making coffee with a paper filter. Water flows through the coffee grounds and the filter paper. The grounds (rejected particles) accumulate on the filter surface, and the coffee (filtrate) drips out below. As more coffee is made, the layer of grounds builds up, slowing down the flow.

Characteristics of DEF:

- o High initial flux that rapidly decreases as the filter cake builds up.
- o Essentially 100% recovery of the fluid phase (minus retained moisture in the cake).
- o Requires periodic backwashing, cleaning, or replacement of the filter medium.
- o Suitable for solutions with very low concentrations of suspended solids

Examples: Cartridge filters (used as pre-filters for RO), syringe filters in labs, coffee filters, domestic water pitcher filters.

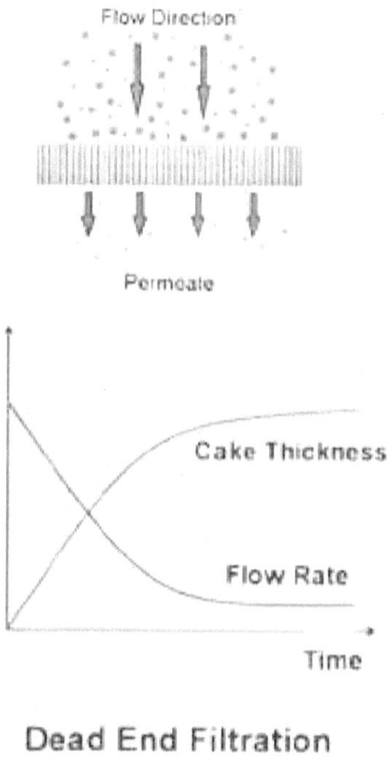

Figure 2.1: Dead-End Filtration

Description: Feed flows directly towards the filter. Solids accumulate on the filter surface, eventually blocking or "blinding" it, requiring cleaning or replacement.

.DEF is **not suitable for processes** like RO where the feed water contains significant concentrations of dissolved solids (salts) and potentially suspended solids or colloids. The rejected salts would rapidly concentrate at the membrane surface, dramatically increasing osmotic pressure locally (a phenomenon called concentration polarization, discussed in Chapter 3) and quickly stopping water flow.

Precipitated salts (scale) and accumulated particles would quickly foul the membrane.

2.3.2 Cross-Flow Filtration: Sweeping the Membrane Clean

To overcome the limitations of DEF, Reverse Osmosis (along with other membrane processes like Nanofiltration, Ultrafiltration, and Microfiltration) employs Cross-Flow Filtration (CFF), also known as Tangential Flow Filtration (TFF).

In CFF, the feed water flows parallel or tangentially across the membrane surface at a relatively high velocity. Hydraulic pressure still forces some of the water through the membrane (as permeate), but the sweeping action of the parallel flow helps to keep rejected particles and concentrated solutes from accumulating directly on the membrane surface.

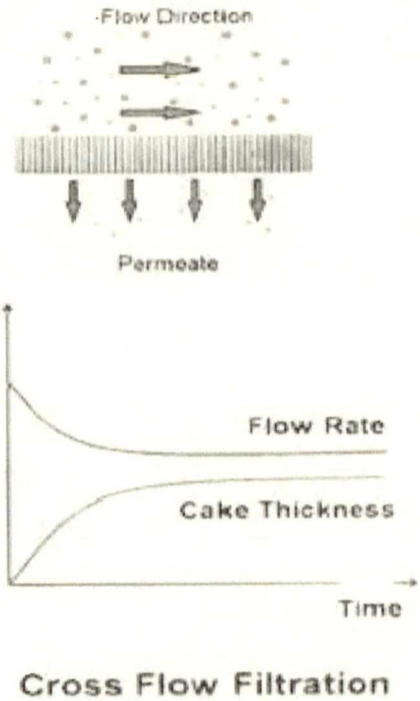

Figure 2.2: Cross-Flow Filtration

Analogy: Imagine trying to wash fine sand off a slightly porous sidewalk. If you just poured water straight down (like DEF), the sand would quickly clog the pores. But if you sprayed the water at an angle, sweeping it across the surface (like CFF), the flow would carry most of the sand away, while some clean water would still seep through the pores.

Description: Feed flows parallel to the membrane surface. Pressure drives water through the membrane (permeate). The cross-flow sweeps away rejected solutes and particles, exiting as a concentrated stream. A boundary layer with higher solute concentration still forms near the membrane surface.

Characteristics of CFF:

- Maintains a more stable flux over time compared to DEF because fouling is reduced (though not eliminated).
- Generates two output streams:
  - **Permeate:** The purified water that has passed through the membrane.
  - **Concentrate (or Reject/Brine)**: The remaining feed flow containing the concentrated rejected solutes and particles that were swept along the membrane surface.
- Requires energy not only to pressurize the feed for permeation but also to maintain the cross-flow velocity (recirculation or high feed flow).
- Allows for much higher water recovery rates than would be possible with DEF in RO applications.
- The cross-flow velocity is a critical design parameter. It needs to be high enough to create sufficient turbulence and shear force at the membrane surface to minimize the buildup of rejected solutes (control concentration polarization) and prevent particles from settling. However, excessively high velocity increases pressure drop along the flow channel and consumes more energy. Finding the optimal balance is key to efficient RO system design and operation.
- All commercial RO systems operate in cross-flow mode, utilizing module designs (like spiral-wound or hollow fiber,

discussed in Chapter 4) that facilitate this tangential flow across large membrane areas packed into compact volumes.

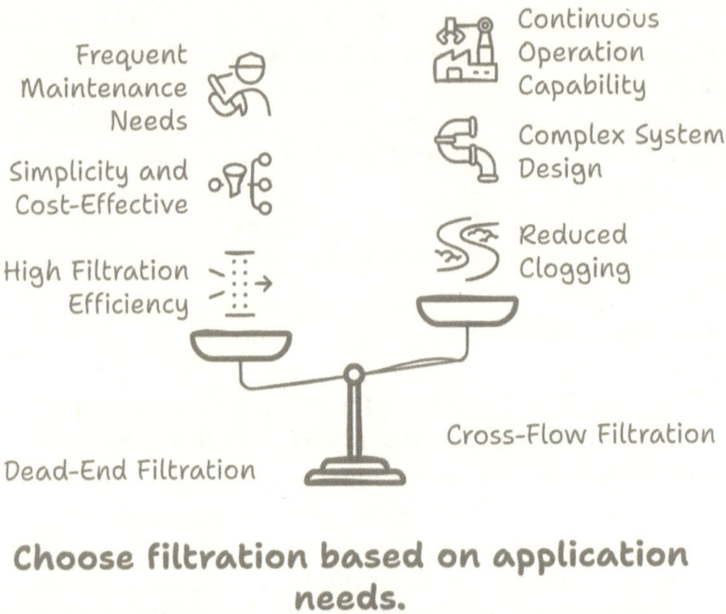

Figure 2.3: Dead-End Filtration vs Cross-Flow Filtration

2.4 Chapter Summary and Key Takeaways

- Osmosis is the natural movement of water across a semi-permeable membrane from a dilute solution to a concentrated solution, driven by differences in water potential.
- Osmotic Pressure (π) quantifies this tendency and must be overcome in RO. It increases with solute concentration and temperature, and can be estimated using the Van't Hoff equation or approximated using TDS levels.

- Reverse Osmosis uses applied hydraulic pressure greater than the osmotic pressure to force water flow in the reverse direction (from concentrated to dilute).
- The Net Driving Pressure (NDP) is the effective pressure causing water permeation (NDP = Pfeed - Ppermeate - $\Delta\pi$).
- Water flux through the membrane is primarily driven by NDP, while salt flux is mainly driven by the concentration difference. Higher pressure generally improves both water production and quality.
- Temperature significantly impacts water flux (higher temp = higher flux) and slightly impacts salt flux.
- RO utilizes Cross-Flow Filtration (CFF), where feed water sweeps tangentially across the membrane surface. This minimizes the buildup of rejected solutes and particles, allowing for sustained operation, unlike Dead-End Filtration (DEF).
- CFF produces two streams: purified permeate and a concentrate stream containing the rejected substances.

2.5 References

1. Van't Hoff, J. H. (1887). The role of osmotic pressure in the analogy between solutions and gases. Zeitschrift für Physikalische Chemie, 1, 481-508. (Original work establishing the osmotic pressure law).
2. Mulder, M. (1996). Basic Principles of Membrane Technology (2nd ed.). Kluwer Academic Publishers. (A classic text covering fundamental transport phenomena in membranes).
3. Baker, R. W. (2012). Membrane Technology and Applications (3rd ed.). John Wiley & Sons. (Chapters on RO fundamentals and transport).
4. AWWA Manual M46: Reverse Osmosis and Nanofiltration (2007). American Water Works Association. (Provides practical explanations of principles relevant to water treatment).
5. Perry, R. H., & Green, D. W. (Eds.). (2008). Perry's Chemical Engineers' Handbook (8th ed.). McGraw-Hill. (Section on

Membrane Separation Processes contains detailed theoretical background).

6. Hoek, E. M. V., & Elimelech, M. (2003). Cake-enhanced concentration polarization: a new fouling mechanism for salt-rejecting membranes. Environmental Science & Technology, 37(24), 5581–5588. (Discusses the interaction between particle fouling and concentration polarization).

Chapter 3: The RO Vocabulary - Following the Water's Journey

3.1 Introduction: Why We Need a Common Language

Imagine trying to build a car engine with mechanics who all use different names for the pistons, cylinders, and spark plugs. It would be chaos! Similarly, the world of Reverse Osmosis has its own specialized vocabulary. To design, operate, troubleshoot, or even just intelligently discuss an RO system, we need a shared understanding of the fundamental terms and metrics.

In the previous chapters, we explored the basic 'what' and 'why' of RO and the scientific principles behind it. Now, we'll equip ourselves with the practical language used every day in the field. We'll do this by following a parcel of water as it journeys from the feed tank, through the pretreatment steps, across the membrane, and finally exits as either purified product or concentrated reject. Along this path, we'll define the key terms that measure the challenges, the performance, and the potential pitfalls of the process.

3.2 Starting Point: Characterizing the Feed Water Challenge

Before we even think about applying pressure, we must understand the raw material: the feed water. Is it seawater teeming with salt? Brackish groundwater laden with hardness minerals? Or surface water carrying fine silts and organic matter? The nature of the feed water dictates the entire design and operational strategy. Two key indices help us assess potential problems before the water hits the delicate RO membranes.

3.2.1 Will it Clog? Assessing Particulate Risk with Silt Density Index (SDI)

RO membranes have incredibly fine pores, designed to block dissolved salts but allow water through. However, they are easily choked by microscopic suspended particles – silt, clay, organic debris, corrosion products. Imagine trying to pour slightly muddy water through a coffee filter; it quickly slows to a trickle as the fine particles block the pores.

The Silt Density Index (SDI) is our standard way to gauge this particulate plugging potential. It's not a count of particles, but an empirical measure of how quickly a standard 0.45-micron filter paper gets plugged when the feed water is forced through it under controlled pressure (30 psi).

- **Calculation Recap:** SDI(T_total) = [1 - (ti / tf)] * 100 / Ttotal (where ti and tf are initial and final times to filter a set volume, and Ttotal is usually 15 minutes).
- **The Goal:** For the long, thin channels and sensitive surfaces inside RO elements, we need very clean water. Membrane manufacturers almost universally require an SDI15 value below 5, and strongly recommend below 3, for their warranties to be valid.
- **The Implication**: A high SDI value (>5) tells us the pretreatment system (media filters, ultrafiltration, etc.) is either inadequate or not performing correctly. It's a red flag signaling an imminent risk of particulate fouling (see Section 3.6.1) on the RO membranes if not addressed. Measuring SDI regularly after pretreatment is a critical check on the health of the system's first line of defense.

3.2.2 Will it Solidify? Predicting Mineral Scale with Langelier Saturation Index (LSI)

Besides suspended particles, feed water contains dissolved minerals – calcium, magnesium, bicarbonate, sulfate, silica, etc. These are the very substances RO is designed to remove. However, as RO removes pure water, these dissolved minerals become increasingly concentrated in the remaining water. Just like dissolving too much sugar in iced tea makes it gritty, concentrating these minerals too much can cause them to precipitate out of solution and form hard, rock-like deposits called scale on the membrane surface.

The most common culprit, especially in non-seawater applications, is calcium carbonate ($CaCO_3$) – essentially limestone. The Langelier Saturation Index (LSI) is our predictive tool for $CaCO_3$ scale. It compares the water's actual pH to the calculated pH at which it would be perfectly saturated with $CaCO_3$ (pHs).

- **Calculation Recap:** LSI = pH actual - pHs (where pHs depends on Calcium, Alkalinity, TDS, and Temperature).
- Interpretation:
    - LSI > 0: Scaling potential exists (water is supersaturated).
    - LSI = 0: Water is stable (saturated).
    - LSI < 0: Water is potentially corrosive (undersaturated).
- **The Catch**: We don't just calculate LSI for the feed water. We must calculate it for the concentrate stream (see Section 3.5.4), where minerals are most concentrated. An acceptable LSI in the feed can easily become a high positive (scaling) LSI in the concentrate.
- **The Strategy**: Based on the predicted concentrate LSI, we decide on pretreatment: either add acid to lower the pH (making LSI negative) or, more commonly, add an **antiscalant chemical**. Antiscalants act like traffic cops for mineral ions, interfering with crystal formation and allowing the water to remain supersaturated without forming scale. RO design software performs these calculations to determine the maximum safe operating limits and necessary chemical treatment. (Note: For other potential scales like calcium sulfate or silica, different indices or calculations are used, but the principle of predicting precipitation based on concentration is the same).

3.3 Applying the Force: Pressure and the Drive to Purify

Once we've assessed the feed water and hopefully pre-treated it to minimize immediate clogging or scaling risks, we need to provide the driving force for purification. This comes from a high-pressure pump.

3.3.1 Overcoming Resistance: The Role of Net Driving Pressure (NDP)

As we learned in Chapter 2, simply applying pressure isn't enough. We need to overcome the natural osmotic pressure (π) of the feed water, which tries to push water back across the membrane. The actual pressure available to push water in the desired direction (from salty feed to pure permeate) is the Net Driving Pressure (NDP).

- **Formula Recap:** NDP $\approx$ P feed – P permeate – π average_feed
- **The Significance**: NDP is the engine of the RO process. Higher NDP means more driving force, which generally leads to more water production (higher flux, see below). However, osmotic pressure (π) increases as salts get concentrated along the membrane's path, meaning NDP naturally decreases from the beginning to the end of the RO system. The pump must provide enough initial pressure to ensure sufficient NDP exists even at the very end of the membrane train to achieve the desired overall water production. Too little NDP, and water stops flowing through the membrane.

3.4 At the Gatekeeper: Performance at the Membrane Surface

The heart of the RO system is the membrane element itself, typically a spiral-wound configuration (details in Chapter 4) containing large areas of the semi-permeable membrane sheet. Here, the crucial separation happens, governed by two key phenomena:

3.4.1 The Rate of Passage: Membrane Flux (Jw)

How fast does the purified water actually pass through the membrane? That's what Flux tells us. It's not the total system output, but the output rate per unit area of the membrane.

- **Definition:** Volume of permeate per unit membrane area per unit time.
- **Units:** Gallons per square foot per day (GFD) or Liters per square meter per hour (LMH).
- **Analogy:** Think of rain soaking into the ground. Flux is like measuring how many inches of rain soak in per hour in a specific square foot of ground. Different soil types (membranes) and different amounts of rain pressure (NDP) yield different flux rates.
- **Importance:** Flux is directly related to NDP (higher NDP ≈ higher Flux). Designers choose a target average flux based on feed water quality and membrane type. Using a higher flux allows for a smaller system (less membrane area needed) but comes at a cost: it intensifies the 'traffic jam' effect described next. Typical design fluxes range from 8-12 GFD for seawater to 14-18 GFD for brackish water, often lower for difficult-to-treat wastewater. Monitoring changes in flux (corrected for pressure and temperature) is a key way operators detect membrane fouling or scaling.

3.4.2 The Traffic Jam Effect: Concentration Polarization (CP)

This is a critical concept. As water flows through the membrane, the rejected salts get left behind right at the surface. While the cross-flow (water flowing along the membrane) tries to sweep these salts away, it can't keep up perfectly. This results in a thin layer of water directly against the membrane surface that is significantly more concentrated in salts than the bulk feed water flowing past. This phenomenon is Concentration Polarization (CP).

- **Analogy:** Imagine a busy highway exit ramp (water permeating). Cars (salts) that don't take the exit get backed up near the exit point, even if the main highway (bulk flow) is still moving. This local congestion is CP.

- **The Problems:**
  1. **Higher Local Osmotic Pressure**: The increased salt concentration (Cw) at the wall means the osmotic pressure the system must overcome (πw) is higher than that of the bulk feed (πf). This reduces the effective NDP and lowers flux.
  2. **Increased Salt Leakage**: The driving force for salts to leak through the membrane depends on the concentration difference (Cw - Cp). Higher CP increases this difference, leading to slightly more salt passage and poorer permeate quality (lower rejection).
  3. **Enhanced Scaling/Fouling:** This highly concentrated layer is where scaling is most likely to begin, as solubility limits are exceeded first right at the membrane surface. It also provides a rich environment for biofilms to establish.
- **Control:** CP is managed by ensuring adequate cross-flow velocity (sweeping effect) and by not operating at excessively high flux rates (reducing the arrival rate of salts). Feed spacers within the membrane elements also help create turbulence to minimize this boundary layer.

Concentration Polarization Boundary Layer: Components and Implications

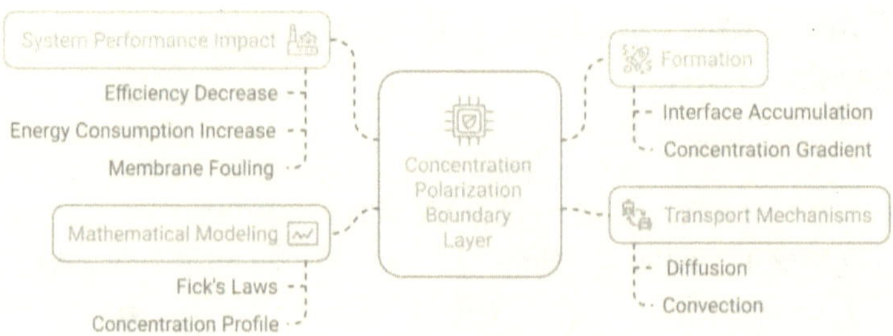

Figure 3.1: Concentration Polarization Boundary Layer: Components and Implications

3.5 The Results: Measuring Outputs and Efficiency

After the water passes (or doesn't pass) through the membrane, we look at the results. What did we produce, and how efficiently did we do it?

3.5.1 How Much Clean Water? System Flow Rating

This is the primary measure of an RO system's capacity – the total volume of purified permeate (product water) it's designed to deliver.

- **Units:** Gallons per day (GPD), cubic meters per hour (m^3/hr), etc.
- **Significance:** This is what the end-user needs. An RO system is specified and sold based on its rated permeate flow under defined conditions (e.g., "a 100,000 GPD brackish water RO system"). The actual operating flow may vary based on temperature, fouling, and pressure.

3.5.2 How Efficiently? Recovery Rate (%)

In cross-flow RO, not all feed water becomes permeate; some must continuously flush salts away as concentrate. Recovery tells us the percentage of incoming feed water that is successfully converted into usable permeate.

- **Calculation:** Recovery (%) = (Permeate Flow Rate / Feed Flow Rate) * 100%
- **Significance:** Higher recovery means less water wasted as concentrate, saving water and reducing disposal volume. However, high recovery leads directly to much higher concentrations in the reject stream (see Beta), increasing scaling potential and osmotic pressure. There's always a trade-off.
- **Typical Ranges:** 40-50% for SWRO, 50-85% for BWRO, often limited by scaling potential of the feed water.

3.5.3 How Pure? Salt Rejection (%)

This metric quantifies how effectively the membrane removed the dissolved salts (or a specific contaminant).

- **Calculation:** Rejection (%) = [1 - (Permeate Concentration / Feed Concentration)] * 100% (using TDS, conductivity, or specific ion concentration).
- **Significance**: Measures the quality of the product water. Modern membranes boast high rejection rates (often >99% for common salts). Rejection can be affected by pressure, temperature, recovery, and membrane condition. A drop in rejection often signals a problem (e.g., membrane damage, leak, fouling).

3.5.4 How Concentrated is the Waste? Beta (Concentration Factor)

While recovery tells us the ratio of flows, the Concentration Factor (Beta, β) tells us how many times more concentrated the salts are in the bulk concentrate (reject) stream compared to the initial feed water.

- **Approximation**: $\beta \approx 1 / (1 - \text{Recovery})$ (where Recovery is a fraction, e.g., 0.75 for 75%).
- **Significance:** This factor is crucial for predicting scaling. If feed water contains 100 ppm of calcium and the system runs at 75% recovery ($\beta \approx 4$), the concentrate will contain roughly 400 ppm of calcium. This value, along with other ion concentrations adjusted by Beta, is plugged into LSI or other scaling calculations to determine if precipitation is likely in the final stages of the RO system.

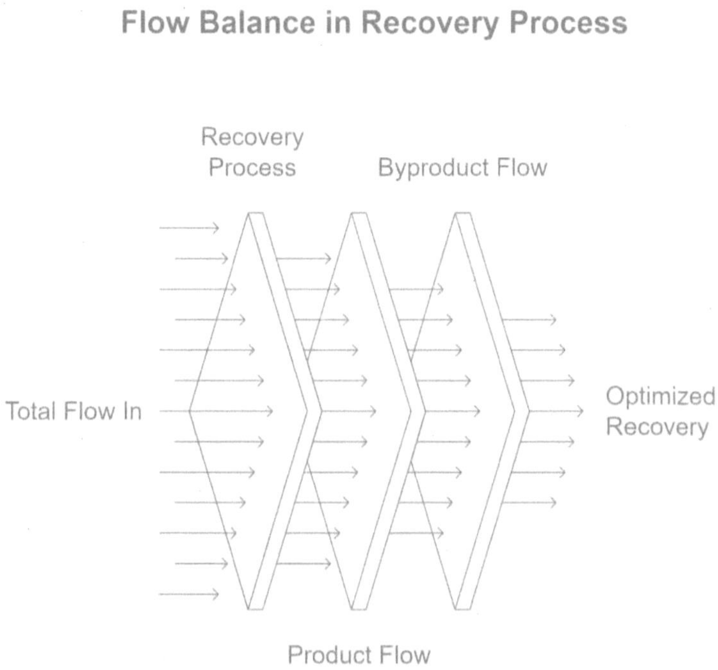

Figure 3.2:Flow Balance in Recovery Process

3.6 The Unwanted Consequences: When Things Go Wrong

Even with careful design and operation, membranes are susceptible to losing performance over time due to the accumulation of unwanted materials.

3.6.1 General Gunk: Understanding Fouling

Fouling is the catch-all term for any process where constituents in the feed water deposit onto the membrane surface or block its pores, hindering performance.

- **Types:** Includes particulate/colloidal fouling (from silt, clay – predicted by SDI), microbial fouling or biofouling (slimy bacterial growth), and organic fouling (from natural organic matter, oils).
- **Symptoms:** Usually causes a decrease in permeate flux (less water production) and/or an increase in the pressure drop across the membrane element (takes more pressure to push water through the fouled channels).
- **Control:** Primarily through effective pretreatment, but also requires periodic chemical cleaning.

3.6.2 Mineral Deposits: The Problem of Scaling

Scaling is a specific, often more stubborn, type of fouling caused by the precipitation of dissolved inorganic minerals when they become too concentrated, exceeding their solubility limit.

- **Common Culprits**: Calcium carbonate ($CaCO_3$ - predicted by LSI), calcium sulfate ($CaSO_4$), silica (SiO_2), barium/strontium sulfate.
- **Cause**: Driven by high concentration factor (Beta) due to high recovery, often exacerbated by concentration polarization (CP) at the membrane surface.
- **Symptoms:** Similar to fouling (flux loss, increased pressure drop), but scale can sometimes also damage the membrane rejection layer, leading to decreased salt rejection (poorer permeate quality). Scale is often harder to remove than other types of fouling.
- **Control**: Limiting recovery, pH adjustment (for $CaCO_3$), and/or adding specific antiscalant chemicals.

3.7 Chapter Summary and Key Takeaways

Understanding RO requires mastering its vocabulary. We've journeyed with the water:

- **Feed Assessment:** We check SDI for particulate risk and LSI (and others) for potential scaling.
- **Driving Force**: The pump provides pressure, but the work is done by the Net Driving Pressure (NDP) overcoming osmotic pressure.
- **Membrane Performance**: Water passes through at a certain Flux, but Concentration Polarization (CP) creates a high-salt layer right at the surface, impacting performance.
- **Outputs**: We measure the total Flow Rating (permeate), the water efficiency (Recovery), the purity (Rejection), and the concentration of the reject (Beta factor).
- **Problems**: We constantly battle Fouling (general gunk) and Scaling (mineral precipitation), which degrade performance and require careful management through pretreatment and cleaning.

These terms are interconnected and essential for interpreting RO system data, diagnosing problems, and ensuring efficient, reliable operation.

3.8 References

1. AWWA Manual M46: Reverse Osmosis and Nanofiltration (2007). American Water Works Association. (Chapters covering terminology, performance metrics, and pretreatment indices).
2. DuPont Water Solutions. (Various Technical Manuals and Literature). e.g., FilmTec™ Membranes Technical Manual, RO & NF Element Performance Normalization. (Provide practical definitions and calculation methods).
3. Hydranautics (A Nitto Group Company). (Technical Service Bulletins and Software Manuals). e.g., IMSDesign Software Help. (Offer definitions and context for design parameters).

4. Langelier, W. F. (1936). The Analytical Control of Anti-Corrosion Water Treatment. Journal of the American Water Works Association, 28(10), 1500–1521. (The original paper defining the LSI).
5. ASTM Standard D4189-07(2014). Standard Test Method for Silt Density Index (SDI) of Water. ASTM International. (Official standard detailing the SDI test procedure).
6. Stiff, H. A., & Davis, L. E. (1952). A Method for Predicting the Tendency of Oil Field Waters to Deposit Calcium Sulfate.

Chapter 4: Membranes - The Heart of Separation

4.1 Introduction: The Semipermeable Barrier

At the very core of every Reverse Osmosis system lies the membrane. It's the gatekeeper, the selectively permeable barrier that performs the 'magic' of separating pure water from dissolved salts and other contaminants. While we often talk about the pumps, the pretreatment, and the overall system, it's the membrane's intrinsic properties that ultimately define the performance – how much water is produced (flux), how pure it is (rejection), and how much pressure is needed.

In this chapter, we'll peel back the layers of this critical component. First, we'll explore the theoretical models scientists use to understand and predict how water and solutes move through these dense films. Second, we'll examine the evolution of membrane materials, from the early cellulose acetate pioneers to the sophisticated thin-film composites that dominate today. Finally, we'll look at how these delicate membrane sheets are cleverly packaged into practical, robust modules that can be installed in pressure vessels and handle large volumes of water.

4.2 How Does it Work? Transport Models

Understanding exactly how a membrane separates water from salt at the molecular level is complex and still a subject of scientific refinement. Several models have been proposed, each capturing different aspects of the process and having different assumptions about the membrane's structure (is it truly dense, or does it have tiny pores?).

4.2.1 The Dominant Theory: Solution-Diffusion Model (non-porous)

The most widely accepted and used model for RO (and gas separation) membranes, particularly the dense polyamide type, is the Solution-

Diffusion (S-D) model, first proposed by Lonsdale and others based on earlier work by Graham. This model assumes the membrane material itself is a non-porous, dense phase. Transport occurs in three distinct steps:

1. **Sorption (Solution):** Molecules from the feed solution (both water and solutes) dissolve into the membrane material at the high-pressure feed-side interface, similar to how gas dissolves in a liquid. This step is governed by equilibrium partitioning – different molecules have different solubilities in the membrane polymer.
2. **Diffusion:** Once dissolved in the membrane phase, the molecules move randomly under a concentration gradient (or more accurately, a chemical potential gradient) from the high-concentration (feed) side to the low-concentration (permeate) side. This diffusion rate depends on the molecule's size, shape, interaction with the polymer chains, and the inherent diffusivity (mobility) within the membrane material.

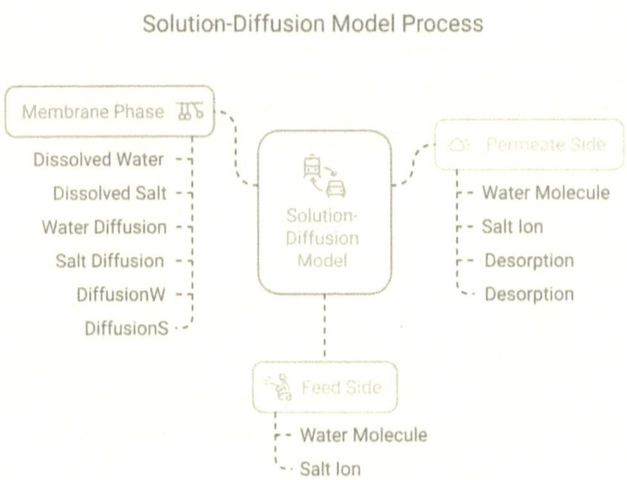

Figure 4.1: Solution-Diffusion Model Process

Description: Solution-Diffusion Model Process.

3. **Desorption:** Molecules reaching the low-pressure permeate-side interface desorb (evaporate or dissolve) out of the membrane phase into the permeate stream.

How does separation occur in the S-D model? It's a combination of differences in solubility and diffusivity:

- **Water:** Small, polar water molecules typically have relatively high solubility and high diffusivity in the hydrophilic polymers used for RO membranes (like polyamide).
- **Salts (Ions)**: Dissolved salts exist as charged ions (e.g., Na^+, Cl^-) surrounded by hydration shells of water molecules. These hydrated ions are generally much less soluble in the dense, relatively low-dielectric polymer matrix compared to water. Furthermore, their larger effective size and charge make their diffusivity within the polymer matrix significantly lower than that of water.

Therefore, water dissolves readily and diffuses relatively quickly, while salts struggle to dissolve and diffuse much more slowly. This difference leads to the selective passage of water over salts.

Key Equations (Conceptual):

- Water Flux (Jw) $\approx$ A * (ΔP - $\Delta \pi$) where 'A' (Water Permeability Coefficient) depends on water's solubility and diffusivity.
- Salt Flux (Js) $\approx$ B * ($C_{f,wall} - C_p$) where 'B' (Salt Permeability Coefficient) depends on salt's solubility and diffusivity.

The S-D model successfully explains many observed phenomena in RO, such as the dependence of water flux on net pressure and the dependence of salt flux primarily on concentration difference.

4.2.2 Accounting for Reality: Solution-Diffusion Imperfection Model (porous hints)

While the S-D model is powerful, researchers recognized that real membranes aren't perfectly dense. They might contain tiny, transient "free volume" gaps or even minute structural defects or "pores." The Solution-Diffusion Imperfection Model modifies the S-D framework by suggesting that a small amount of transport might also occur convectively through these imperfections or pores, alongside the dominant solution-diffusion mechanism. This could help explain why rejection isn't always 100% and why sometimes pressure seems to have a minor influence on salt passage (as convective flow through pores would be pressure-dependent).

4.2.3 Alternative Views: Finely-Porous & Sorption-Capillary Flow Models (porous)

Other models lean more heavily on the idea of the membrane having actual, albeit extremely small, pores.

- **Finely-Porous Model:** This model views the membrane as having nano-scale pores. Separation occurs primarily due to size exclusion (steric hindrance – ions are physically too large to enter or pass through the pores easily) and charge exclusion (Donnan exclusion – if the pore walls carry a fixed charge, they repel ions of the same charge and attract ions of the opposite charge, affecting transport). Water moves through these pores by viscous flow (convection), driven by the pressure gradient.
- **Preferential Sorption-Capillary Flow Model:** This model focuses on the interactions at the membrane surface and within the pores. It proposes that water is preferentially adsorbed onto the pore walls (due to hydrogen bonding or other interactions), forming a layer that flows through the capillary-like pores under pressure. Salts are less readily adsorbed and are thus excluded from this flow path.

The distinction between a "dense" polymer with transient free volume (S-D) and a "porous" polymer with nano-scale pores is subtle and perhaps semantic at the molecular level. Most modern TFC membranes likely exhibit behavior that incorporates elements of both solution-diffusion and some form of restricted flow through intermolecular pathways or "pores."

4.2.4 The Black Box Approach: Irreversible Thermodynamics

A different way to describe transport, without making specific assumptions about the membrane structure (pores vs. dense), comes from the field of irreversible thermodynamics. This approach treats the membrane as a "black box" system where fluxes (of water and solute) are related to the driving forces (pressure difference, concentration difference) through a set of phenomenological coefficients.

The Kedem-Katchalsky model is a well-known example, using three coefficients:

- **Lp**: Hydraulic permeability (related to water flux driven by pressure).
- **σ (Sigma)**: Reflection coefficient (measures the membrane's inherent selectivity; $\sigma=1$ for a perfectly rejecting membrane, $\sigma=0$ for a non-selective one).
- **ω (Omega)**: Solute permeability (related to salt flux driven by concentration).

Flux Equations (Conceptual):

- Volume Flux (Jv) $\approx Lp * (\Delta P - \sigma\Delta\pi)$
- Solute Flux (Js) $\approx \omega\Delta C + (1 - \sigma) * Jv * C\ avg$

This approach provides a mathematically rigorous framework and can describe coupled transport phenomena (where the flow of water can drag solute, and vice versa), but determining the coefficients experimentally can be challenging, and it offers less intuitive physical insight into the membrane structure compared to mechanistic models like S-D.

4.2.5 Why Models Matter

Understanding these models isn't just an academic exercise. They help us:

- **Interpret Experimental Data**: Relate observed flux and rejection to fundamental membrane properties.
- **Predict Performance**: Estimate how a membrane will perform under different operating conditions (pressure, concentration, temperature).
- **Guide Material Science**: Suggest what polymer properties (solubility parameters, free volume, charge) should be targeted to create better membranes with higher flux, higher rejection, or specific selectivity.
- **Develop Design Software**: Form the basis for the complex algorithms used in RO system design software.

While the S-D model remains the practical cornerstone, acknowledging the potential role of pore-flow mechanisms and the insights from irreversible thermodynamics provides a more complete picture.

4.3 What Are They Made Of? Membrane Materials

The performance of an RO membrane is intrinsically linked to the material used to create that thin, selective layer. The ideal material would offer high water permeability, near-perfect salt rejection, excellent mechanical strength, broad chemical resistance (especially to pH extremes and oxidants like chlorine), resistance to fouling, long operational life, and low manufacturing cost. Needless to say, finding a single material that excels in all areas is the ongoing challenge of membrane science.

Table 4.1: Comparison of Major RO Transport Models

Model Name	Core Assumption about Membrane Structure	Primary Separation Mechanism(s)	Key Concepts/Parameters	Main Applicability
Solution-Diffusion (S-D)	Dense, non-porous polymer matrix	Differences in Solubility & Diffusivity	Solubility (K), Diffusivity (D), A & B coeff.	Dense RO/NF, Gas Sep.
S-D Imperfection	Mostly dense + minor pores/defects	S-D + minor Convection through pores	S-D parameters + pore flow term	Modifying S-D for reality
Finely-Porous	Defined nano-scale pores	Size Exclusion (Steric), Charge Exclusion (Donnan)	Pore size, Pore charge, Viscous flow	Porous MF/UF, some NF/RO
Preferential Sorption-Capillary Flow	Porous, with surface interactions	Preferential sorption of water, Capillary flow	Adsorption isotherms, Capillary pressure	Alternative porous view
Irreversible Thermodynamics	"Black Box" (no structural assumption)	Phenomenological coupling of fluxes & forces	L_p (Permeability), σ (Reflection), ω (Solute Perm.)	General, theoretical

4.3.1 The Pioneer: Cellulose Acetate (CA) Asymmetric Membranes

The story of practical RO began with Cellulose Acetate (CA). Derived from natural cellulose (wood pulp or cotton linters), CA was the material used by Loeb and Sourirajan in the early 1960s to create the first high-flux asymmetric membranes.

Structure: Asymmetric means the structure varies across its thickness. They developed a casting process that produced a membrane with a very thin (≈0.1-0.5 μm), dense "skin" layer responsible for the salt rejection, seamlessly integrated with a much thicker (≈100-200 μm), highly porous substructure that provided mechanical support while offering minimal resistance to water flow. This clever structure was the key to achieving practical flux rates.

Advantages:

- **Some Chlorine Tolerance:** CA membranes can tolerate low levels of continuous free chlorine exposure (typically 0.5-1.0 ppm), which simplifies pretreatment compared to chlorine-intolerant membranes.
- **Hydrophilic:** Cellulose acetate is relatively hydrophilic (water-loving), which can help reduce fouling by hydrophobic organic substances.

Disadvantages:

- **pH Sensitivity:** Highly susceptible to hydrolysis (chemical breakdown by water) outside a narrow pH range (typically pH 4-6). Operation at higher or lower pH rapidly degrades the membrane.
- **Biodegradation:** Being derived from cellulose, CA is susceptible to attack by microorganisms, requiring careful biocide management.
- **Compaction:** Under prolonged high pressure, the membrane structure can compact, leading to reduced flux over time.
- **Lower Performance:** Generally exhibit lower salt rejection and lower water flux compared to modern TFC membranes, especially for challenging feeds like seawater.

Current Status: While largely superseded by TFC membranes for most mainstream applications (especially desalination and high-purity water), CA membranes are still used in some specific areas, partly due to their chlorine tolerance, such as certain food processing applications or niche industrial streams. Blends (like Cellulose Triacetate, CTA, or blends with other polymers) were also developed to improve properties.

4.3.2 The Workhorse: Polyamide Thin-Film Composite (TFC) Membranes

The development of Thin-Film Composite (TFC) membranes in the 1970s, primarily using aromatic polyamides for the separation layer, revolutionized the RO industry and remains the dominant technology today.

4.3.2.1 A Quick Look Back: Linear Aromatic Polyamides:

Early research explored homogeneous membranes made from linear aromatic polyamides (like Nomex®-type materials). While offering better chemical stability than CA, their flux was often limited.

4.3.2.2 The Standard: Interfacially Polymerized TFC Polyamides:

The breakthrough was the composite approach combined with interfacial polymerization, pioneered by John Cadotte. This method allows for independent optimization of the membrane layers:

1. Support Layer (Microporous): A pre-formed, robust microporous layer, typically made of polysulfone (PSU) or polyethersulfone (PES), provides the mechanical strength and porous pathway for water. This layer itself offers negligible salt rejection. Its pore size is usually in the ultrafiltration range (e.g., 10-50 nm).
2. Fabric Backing: The polysulfone layer is cast onto a non-woven polyester fabric web (like a stiff felt) that provides overall mechanical handling strength and prevents the membrane from extruding through support structures in the module.
3. Separation Layer (Ultrathin Polyamide): The magic happens here. An extremely thin (typically 0.1-0.2 μm, i.e., 100-200 nanometers!), highly cross-linked aromatic polyamide layer is formed in-situ directly on top of the polysulfone support layer

via interfacial polymerization. This involves contacting the PSU surface (soaked with an aqueous amine monomer, typically m-phenylenediamine, MPD) with an organic solvent solution containing a reactive acyl chloride monomer (typically trimesoyl chloride, TMC). The polymerization reaction occurs extremely rapidly right at the interface between the two immiscible solutions, forming the ultra-thin, highly selective polyamide skin.

Advantages of TFC Polyamides:

- **High Flux and Rejection:** The ultrathin separation layer minimizes resistance to water flow, allowing high flux. The specific chemistry of the cross-linked aromatic polyamide provides excellent intrinsic salt rejection (>99% for NaCl is common).
- **Wide pH Tolerance**: Stable over a much broader pH range (typically pH 2-11) compared to CA, allowing more flexibility in operation and cleaning.
- **Good Mechanical Strength:** The composite structure provides durability.
- **Biological Stability:** Not readily biodegradable like CA.

The Achilles' Heel:

- **Chlorine Intolerance:** This is the major drawback. The amide linkages (-NH-CO-) in the polyamide structure are susceptible to attack by strong oxidizers like free chlorine (hypochlorite). Chlorine exposure irreversibly damages the membrane, destroying its salt rejection capability. This necessitates careful dechlorination of the feed water (e.g., using sodium bisulfite injection or activated carbon filters) before it reaches the TFC membranes. Even trace amounts of continuous chlorine can cause damage over time. Tolerance to chloramines is slightly better but still limited.

4.3.3 Making Better Membranes: Improvements to TFC Polyamides

While the basic TFC structure persists, significant research focuses on refining the polyamide layer and support structure to enhance performance and overcome limitations:

- **Surface Modifications for Fouling Resistance**: This is a major area. Strategies include:
    - **Coating/Grafting:** Applying thin hydrophilic coatings or grafting hydrophilic polymer chains onto the surface to repel hydrophobic foulants and reduce protein/bacterial adhesion.
    - **Surface Charge Modification:** Adjusting the surface charge (polyamide is typically negatively charged at neutral pH) to repel negatively charged foulants or control interaction with ions.
    - **Surface Smoothing:** Creating smoother surfaces at the nanoscale to reduce sites where foulants can attach and accumulate.
- **Nanotechnology Integration:** Incorporating nanomaterials into the polyamide active layer or the polysulfone support:
    - Carbon Nanotubes (CNTs), Graphene Oxide (GO): Explored to potentially create channels for enhanced water permeability or improve mechanical strength.
    - Zeolites, Metal-Organic Frameworks (MOFs): Nanoparticles with well-defined pore structures incorporated to potentially improve permeability and selectivity.
    - Silver Nanoparticles: Added to impart antimicrobial properties and reduce biofouling (concerns about silver release exist).

(Note: Many nanomaterial approaches are still primarily in research or early commercial stages, facing challenges in scalability, long-term stability, and cost).

- **Optimized Polyamide Chemistry**: Tailoring the amine and acyl chloride monomers or reaction conditions to fine-tune the cross-linking density, free volume, and charge characteristics of the polyamide layer to achieve:

- Higher Rejection: Especially for difficult-to-remove small, neutral molecules like boron (critical for agriculture using desalinated water).
- Higher Flux / Lower Energy: Achieving the same water throughput at lower pressure.
- Improved Chlorine Tolerance: The 'holy grail' for TFCs. Some incremental improvements in tolerance to occasional or low-level exposure have been claimed, but a truly chlorine-robust high-performance RO membrane remains elusive.

- **Support Layer Enhancements:** Modifying the polysulfone support for better porosity, hydrophilicity, or mechanical properties.

4.3.4 Beyond Polyamide: Other Membrane Materials

While polyamide TFCs dominate, research explores other polymers and materials:

- Polyvinyl Alcohol (PVA), Polybenzimidazole (PBI), Sulfonated Polymers (e.g., SPES): Explored for potential advantages in specific areas like chlorine tolerance or solvent resistance, but often face trade-offs in flux/rejection performance for aqueous RO.
- Ceramic Membranes: Made from inorganic materials like alumina, titania, or zirconia. Offer exceptional chemical, thermal, and mechanical stability and inherent fouling resistance. However, they are typically much more expensive to produce in the defect-free forms needed for RO rejection levels and are generally limited to microfiltration/ultrafiltration applications currently, though research into NF/RO ceramics continues.
- Biomimetic Membranes: Incorporating or mimicking biological channels like aquaporins (highly efficient water channels from cell membranes) into synthetic membranes is an exciting research frontier aiming for ultra-high permeability and selectivity.

Table 4.2: Comparison of Cellulose Acetate (CA) and Polyamide TFC Membranes

Feature	Cellulose Acetate (CA) - Asymmetric	Polyamide (PA) - Thin-Film Composite (TFC)	Comments
Structure	Asymmetric (Skin + Porous Support)	Composite (PA Layer + PSU Support + Web)	TFC allows independent layer optimization
Rejection (NaCl)	Good (e.g., 95-98%)	Excellent (e.g., 98% to >99.5%)	TFC generally superior, esp. for SWRO
Flux	Moderate	High	TFC typically 1.5-2x flux of CA at same conditions
pH Tolerance	Narrow (pH 4-6)	Wide (pH 2-11)	Major advantage for TFC
Chlorine Tolerance	Fair (0.5-1.0 ppm free Cl_2)	Very Poor (<0.1 ppm free Cl_2)	Major advantage for CA; Key limitation for TFC
Biodegradability	Susceptible	Resistant	CA requires biocide control
Temperature Limit	Moderate (~35-40°C)	Higher (~45°C)	TFC generally more robust
Primary Use	Niche/Historical	Industry Standard (BWRO, SWRO, etc.)	TFC dominates modern applications

4.4 How Are They Packaged? Membrane Modules

A single square foot or square meter of membrane doesn't process much water on its own. To build practical RO systems capable of handling large flows, vast areas of membrane need to be packaged into manageable units called modules or elements, which can then be loaded into pressure vessels. The module design dictates how efficiently membrane area is packed, how feed water flows across the surface (cross-flow), how permeate is collected, and how resistant the unit is to fouling.

4.4.1 Early Designs: Plate and Frame Modules

One of the earliest designs, adapted from filter presses.

- **Structure:** Consists of flat sheet membranes sandwiched between grooved or channeled support plates, stacked together in a frame. Feed water flows across one side of the membrane, permeate passes through and collects in the channels on the other side.
- **Analogy:** A stack of vinyl records with membranes between them, where grooves collect the permeate.
- **Advantages**: Relatively easy to dismantle for inspection or manual cleaning of individual sheets. Can handle relatively high solids loading compared to spiral wound.
- **Disadvantages**: Very low packing density (membrane area per unit volume of module) – makes systems bulky and expensive for large flows. Numerous seals create potential for leaks. Higher capital cost per unit area.
- **Use:** Largely obsolete for large-scale water RO, but sometimes still found in specific small-scale, high-value applications like pharmaceutical or food processing where ease of cleaning/sanitization is paramount, or for membrane testing.

4.4.2 Handling Tough Feeds: Tubular Modules

Designed for durability and handling difficult feed streams.

- **Structure**: Membranes are cast or coated either on the inside surface of porous support tubes (often 1/2 to 1 inch diameter) or sometimes on the outside. Multiple tubes are bundled together inside a housing (shell). Feed flows through the inside of the tubes (lumen feed) at high velocity, permeate passes through the membrane and tube wall, collecting in the shell. Alternatively, feed can be on the shell side.

- **Advantages:** Excellent resistance to fouling and plugging due to the open channel and high feed flow velocity. Relatively easy to clean, even mechanically (e.g., using sponge balls for internal membrane configurations).
- **Disadvantages:** Extremely low packing density (even lower than plate & frame). High capital cost. High energy consumption due to the pressure drop associated with high velocity flow through tubes.
- **Use:** Primarily for treating highly fouling or viscous industrial effluents, food processing streams (like juice concentration), or oily wastewater, where other module types would fail quickly. Rarely used for typical desalination or potable water production.

4.4.3 The Industry Standard: Spiral Wound Modules (SWMs)

This design, developed in the 1960s, brilliantly solved the problem of packing large membrane areas into compact, cylindrical units, making large-scale RO economical. It is by far the most common module type used today.

- **Structure**: Built around a central perforated permeate collection tube. Large, flat sheet membrane "leaves" are prepared. A leaf consists of two layers of membrane active surface facing outwards, sandwiching a sheet of porous permeate collection spacer material (like a fabric or netting). This sandwich is sealed along three edges (using glue). Several of these leaves, separated by sheets of plastic mesh feed channel spacer (which provides the channel for feed water flow and promotes turbulence), are then glued along their unsealed edge to the central permeate tube and rolled up tightly around it, like a scroll or jelly roll. The entire roll is then wrapped in an outer shell (often fiberglass) for protection and to ensure a snug fit inside the pressure vessel. Standard industrial sizes are typically 8 inches in diameter and 40 inches long (containing 300-450 ft^2 of membrane area), but other sizes exist (e.g., 4-inch, 16-inch diameter).

- **Analogy:** Rolling up layers of window screen (feed spacer) and fabric (permeate spacer with membrane glued on both sides) around a perforated pipe.
- **Flow Paths:**
  1. **Feed Flow:** Pressurized feed water enters one end of the module and flows axially (parallel to the central tube) through the channels created by the feed spacer mesh, across the membrane surfaces.
  2. **Concentrate Flow:** Water that doesn't pass through the membrane continues flowing axially and exits the other end of the module as concentrate.
  3. **Permeate Flow:** Water that passes through the membrane enters the permeate collection spacer within the sealed leaf. Because the leaf is sealed on three sides, this permeate water is forced to flow spirally inward along the permeate spacer channels until it reaches the perforations in the central tube. It then flows down the central tube and exits the module (usually from one or both ends via special seals called interconnectors).
- **Advantages:** Very high packing density. Relatively low manufacturing cost due to automated production. Established technology with proven performance. Wide range of sizes and membrane types available.
- **Disadvantages:** Susceptible to fouling by particulates, which can clog the narrow feed channels (especially if pretreatment is inadequate - hence the importance of low SDI). Channeling (uneven flow distribution) can occur. Cleaning can be effective but it's difficult to ensure uniform cleaning chemical distribution throughout the complex structure. Cannot be mechanically cleaned.

4.4.4 High Packing Density: Hollow Fine Fiber (HFF) Modules

An alternative approach offering even higher packing density.

- **Structure:** Thousands (or even millions) of extremely fine, self-supporting membrane fibers, resembling fine spaghetti or thread (typically < 200 μm outer diameter), are bundled together. The bundle is looped, and the open ends are potted into a resin tube sheet at one end of the module shell. The fibers fill the shell.
- **Flow Paths (Typical configuration - outside-in):** Pressurized feed water enters the shell side and flows radially or cross-flow around the outside of the fibers. Water permeates through the fiber wall into the hollow bore (lumen). The purified permeate flows down the bore to the open ends at the tube sheet and is collected. Concentrate exits from the shell side. (Inside-out flow is also possible but less common for RO).
- **Advantages:** Highest packing density of all module types. Potentially lower fouling in some cases due to absence of spacer mesh (though fibers themselves can foul).
- **Disadvantages**: Very susceptible to plugging by particulates (fibers can easily get blocked). Generally lower pressure tolerance than SWMs. Difficult to clean effectively once plugged. Lower permeate flux per fiber often requires enormous numbers of fibers.
- **Use:** Historically important (e.g., DuPont's original Permasep® B-9 SWRO and B-10 BWRO modules). Still used in some BWRO and gas separation applications, but SWMs have largely taken over for SWRO and most new large-scale water treatment applications due to better fouling management and robustness.

4.4.5 Variations on a Theme: Low-Differential-Pressure Modules

These are typically Spiral Wound Modules designed specifically to minimize the pressure loss (differential pressure, ΔP) between the feed inlet and the concentrate outlet.

- **Design:** Achieved by using thicker feed channel spacers (e.g., 34 mil or even 48 mil instead of the standard 28 or 31 mil). This creates wider channels for feed flow, reducing friction loss.

- **Trade-off:** Thicker spacers mean slightly less membrane area can be packed into the same module volume (lower packing density).
- **Use:** Often used in the later stages of multi-stage RO systems where the feed flow is lower but fouling potential (especially biofouling) might be higher, or in applications treating water with higher fouling tendency (like wastewater reuse) where maintaining good flow distribution and reducing pressure drop is critical.

4.4.6 Variations on a Theme: High-Productivity Modules

Again, typically Spiral Wound Modules, but utilizing membranes specifically engineered for higher water flux (higher 'A' value) at standard operating pressures.

- **Design:** Achieved through modifications to the polyamide layer chemistry or structure during manufacturing.
- **Trade-off:** Often comes at the cost of slightly lower salt rejection compared to standard high-rejection elements.
- **Use**: Suitable for applications where the absolute highest purity is not required, but high water output at lower pressure (saving energy) is desirable. Examples include first-pass SWRO (where permeate will be treated again), certain industrial processes, or high-flow brackish water applications where feed salinity is moderate.

4.4.7 Niche Players: Other Membrane/Module Types

Capillary Modules: Use larger diameter fibers (0.5 - 5 mm) than HFF, often in a configuration similar to tubular but with smaller tubes. Bridge the gap between HFF and tubular. Often used in Ultrafiltration/Microfiltration but sometimes explored for RO/NF.

Vibratory Shear Enhanced Processing (VSEP): A system using intense vibration or rotation of the membrane surface relative to the feed flow to create very high shear rates, minimizing fouling and allowing treatment of very difficult streams. It's more of a system design using modules (often plate-and-frame variants) than a module type itself. High cost and complexity.

Table 4.3: Comparison of Major RO Module Configurations

Feature	Plate & Frame	Tubular	Spiral Wound (SWM)	Hollow Fine Fiber (HFF)
Packing Density	Low	Very Low	**Very High**	**Extremely High**
Fouling Resistance (Particulates)	Moderate	**Excellent**	Moderate	Poor
Ease of Cleaning	Good (Manual Poss.)	Excellent (Mech. Poss.)	Moderate (Chemical)	Difficult
Capital Cost / Area	High	Very High	**Low**	Moderate to Low
Pressure Drop	Moderate	High	Moderate to Low	Low
Typical Use	Niche/Historical	High Fouling Ind.	**Industry Standard**	BWRO/Gas Sep/Historical

4.5 Chapter Summary and Key Takeaways

- The membrane is the selective barrier enabling RO, and its performance depends on material properties and packaging.
- Transport models (like Solution-Diffusion) help explain and predict how water permeates while salts are rejected, based on solubility and diffusivity differences in the membrane material. Porous models consider exclusion mechanisms.
- Membrane materials have evolved from Cellulose Acetate (CA), known for some chlorine tolerance but limited pH range and performance, to the dominant Polyamide Thin-Film Composites (TFCs), offering high flux/rejection and wide pH stability but suffering from chlorine intolerance. Ongoing

research aims to improve TFC fouling resistance and other properties.

- Membrane modules package large membrane areas. While Plate & Frame and Tubular modules exist for niche/difficult applications, the Spiral Wound Module (SWM) is the industry standard due to its high packing density and cost-effectiveness. Hollow Fine Fiber (HFF) offers even higher density but is more susceptible to plugging.
- Specialized SWMs exist, such as low-ΔP variants for fouling control and high-productivity elements for maximizing flow.
- Understanding the interplay between membrane material, transport mechanisms, and module design is crucial for selecting, designing, and operating RO systems effectively.

4.6 References

1. Baker, R. W. (2012). Membrane Technology and Applications (3rd ed.). John Wiley & Sons. (Excellent chapters on RO theory, materials, and modules).
2. Mulder, M. (1996). Basic Principles of Membrane Technology (2nd ed.). Kluwer Academic Publishers. (Classic text on membrane science fundamentals).
3. Lonsdale, H. K., Merten, U., & Riley, R. L. (1965). Transport Properties of Cellulose Acetate Osmotic Membranes. Journal of Applied Polymer Science, 9(4), 1341–1362. (Seminal paper on S-D model for CA).
4. Cadotte, J. E. (1985). Evolution of Composite Reverse Osmosis Membranes. In D. R. Lloyd (Ed.), Materials Science of Synthetic Membranes (ACS Symposium Series, Vol. 269, pp. 273–294). American Chemical Society. (Describes the development of TFC membranes).
5. Petersen, R. J. (1993). Composite reverse osmosis and nanofiltration membranes. Journal of Membrane Science, 83(1), 81–150. (Comprehensive review of TFC membrane chemistry and structure).
6. Elimelech, M., & Phillip, W. A. (2011). The Future of Seawater Desalination: Energy, Technology, and the Environment. Science, 333(6043), 712–717. (Discusses membrane advancements in the context of desalination).

7. AWWA Manual M46: Reverse Osmosis and Nanofiltration (2007). American Water Works Association. (Practical overview of membrane types and modules used in water treatment).
8. Manufacturer Technical Literature (e.g., DuPont Water Solutions, Hydranautics, Toray, LG Chem): Provide detailed specifications and descriptions of their commercial membrane elements (modules).

Chapter 5: Basic Flow Patterns - Assembling the System

5.1 Introduction: Beyond the Single Module

In Chapter 4, we explored the sophisticated design of individual RO membrane modules, like the spiral-wound element. While a single module performs the core separation task, any practical RO system requires many modules working together to produce the desired quantity and quality of water. Simply connecting dozens of modules in parallel to the feed pump isn't efficient or effective.

As water passes through a membrane, permeate is removed, and the remaining feed water becomes more concentrated. This changes the osmotic pressure and reduces the available flow. To manage these changes effectively and optimize both water recovery and system longevity, modules are arranged in specific hydraulic flow patterns. This chapter delves into these fundamental arrangements: how modules are housed in pressure vessels, how these vessels are connected in series (staging) and parallel (trains), and how techniques like concentrate recycle and double-pass operation are used to meet specific performance goals.

5.2 The Housing: Pressure Vessels

The first step in building a system is housing the individual membrane modules. Since RO operates under high pressure (anywhere from 100 psi for low-TDS brackish water to over 1,000 psi for seawater), the modules must be contained within robust Pressure Vessels, sometimes called housings or tubes.

- **Function:** To safely contain multiple RO modules (elements) in series under high operating pressure and direct the flow of feed, concentrate, and permeate.

- **Construction:** Typically cylindrical tubes made from materials resistant to corrosion and pressure. Common materials include:
    - Fiberglass Reinforced Plastic (FRP): Most common for brackish and seawater applications due to excellent corrosion resistance, high strength-to-weight ratio, and relatively smooth inner surface.
    - Stainless Steel (e.g., 304, 316): Used in some industrial, sanitary (food/pharma), or older applications. Can be susceptible to chloride corrosion in certain water chemistries if the wrong alloy is chosen.
    - Other Composites/Plastics: May be used for lower pressure applications.
- **Configuration:**
    - Standard pressure vessels are designed to hold multiple membrane elements end-to-end, typically 1 to 7 (or even 8) elements per vessel. 8-inch diameter vessels holding 6 elements are very common in industrial systems.
    - Each end of the vessel has an end cap or head that seals the vessel and provides ports for feed inlet, concentrate outlet, and permeate outlet connections.
    - Special interconnectors (small plastic or stainless steel couplings with O-rings) connect the central permeate tubes of adjacent membrane elements within the vessel, allowing the permeate to flow from one element to the next and ultimately exit the vessel.
    - Thrust cones or similar devices at the downstream end absorb the axial force exerted by the pressurized flow on the elements.

Loading multiple elements into a single vessel reduces the number of expensive high-pressure connections and simplifies plumbing compared to having individual vessels for each element.

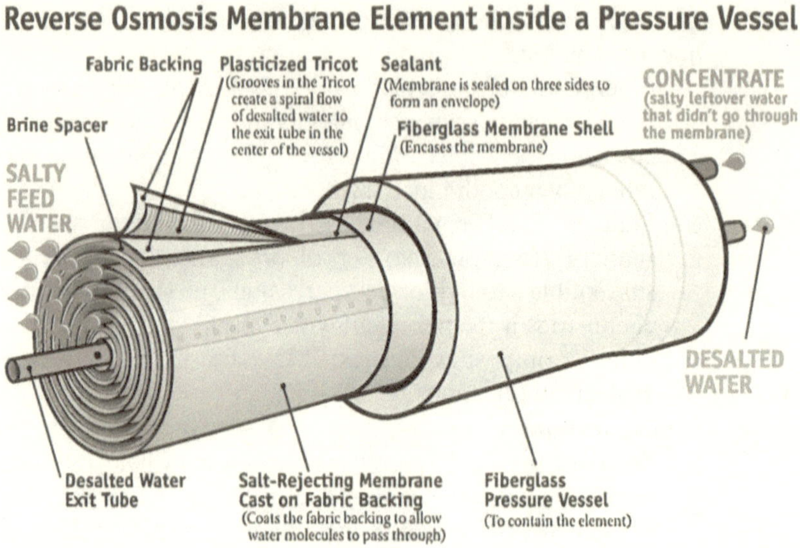

Figure 5.1: Membrane Elements within a Pressure Vessel (Image from Chem1.com)

Description: The efficiency and cost or RO is critically dependent on the properties of the semipermeable membrane.

5.3 Series Flow: Staging in Arrays

Simply connecting multiple pressure vessels in parallel to a feed pump isn't optimal. As water permeates through the membranes in the first few elements, the flow rate of the remaining feed water decreases, and its salt concentration increases. If all vessels received the same initial feed flow, the later elements (or vessels fed in parallel) would operate at very low cross-flow velocities and face highly concentrated feed, leading to severe concentration polarization, fouling, and potentially poor performance.

To counteract this, pressure vessels are commonly arranged in stages, where the concentrate from one set of vessels becomes the feed for the next set. This is known as Concentrate Staging.

5.3.1 The Need for Staging

- **Maintain Cross-Flow Velocity:** As permeate is removed, the total flow decreases. Staging reduces the number of parallel flow paths in subsequent stages, forcing the reduced flow through fewer vessels/elements, thus helping to maintain adequate cross-flow velocity to sweep the membrane surfaces and control CP and fouling.
- **Optimize Recovery:** Allows the system to achieve higher overall recovery by progressively concentrating the feed water across stages, while managing flux and cross-flow within acceptable limits for each stage.
- **Improve Efficiency:** By arranging vessels in stages, designers can tailor the number of elements and vessels in each stage to better match the changing flow rates and pressures, leading to more balanced operation.

5.3.2 Concentrate Staging: The Workhorse Configuration

This is the most common staging method in single-pass RO systems.

How it Works:

1. Feed water enters the first stage, which typically has the largest number of parallel pressure vessels.
2. Permeate is collected from all vessels in the first stage.
3. The concentrate exiting the vessels of the first stage is collected and piped directly as feed into the second stage.
4. The second stage has fewer parallel pressure vessels than the first stage.
5. Permeate is collected from the second stage vessels (often combined with first-stage permeate).

6. The concentrate from the second stage either becomes the final system concentrate (in a two-stage system) or is fed to a third stage (which would have even fewer vessels).

Result: The flow rate decreases from stage to stage, but by reducing the number of parallel vessels, the flow velocity within each vessel's feed channels can be maintained at reasonable levels. The concentration progressively increases through the stages.

5.3.3 Understanding Array Notation (e.g., 2:1, 4:2:1)

A shorthand notation, called an array, is used to describe the number of pressure vessels in each stage.

Notation: X:Y (for a two-stage system) or X:Y:Z (for a three-stage system), where X, Y, and Z represent the number of pressure vessels arranged in parallel within the first, second, and third stages, respectively.

Examples:

- **2:1 Array**: Two pressure vessels in the first stage, their combined concentrate feeds one pressure vessel in the second stage. Total of 3 vessels.
- **4:2:1 Array:** Four vessels in stage 1 feed two vessels in stage 2, which feed one vessel in stage 3. Total of 7 vessels.
- **10:5 Array:** Ten vessels in stage 1 feed five vessels in stage 2. Total of 15 vessels.

The Ratio: The ratio of vessels between stages (e.g., 2:1, or 4:2 = 2:1, or 10:5 = 2:1) is often around 2:1. This ratio helps maintain relatively consistent feed flow per vessel across the stages, as roughly half the water might be removed as permeate in the preceding stage (at 50% stage recovery). However, the optimal ratio depends on the target recovery, membrane properties, and feed water characteristics, and is determined using design software.

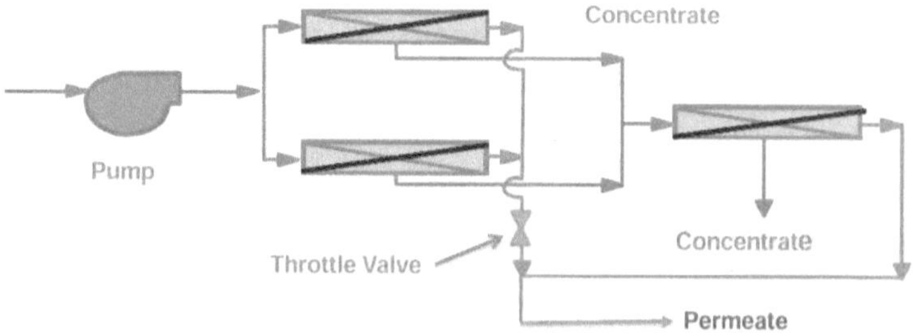

Figure 5.2: Example of a Two-Stage Concentrate Staged Array (2:1 Array)

Description: Feed enters two parallel pressure vessels in Stage 1. Their combined concentrate (Stage2Feed) becomes the feed for one parallel vessels (PV2a-b) in Stage 2. Permeate is collected from both stages. The final concentrate exits Stage 2.

5.3.4 Balancing Flux Across Stages

A key goal in array design is to balance the workload (flux) across the stages and elements.

- **Lead Elements**: The first elements in Stage 1 see the highest feed pressure and lowest osmotic pressure, resulting in the highest NDP and highest flux.
- **Tail Elements:** The last elements in the final stage see the lowest feed pressure (due to pressure drop) and the highest osmotic pressure (due to concentration), resulting in the lowest NDP and lowest flux.
- **Design Limits:** The design must ensure that the flux in the lead elements doesn't exceed the maximum recommended flux (to prevent rapid fouling) and that the flux in the tail elements is still sufficient to contribute meaningfully to production and isn't zero. Design software carefully calculates flows, pressures, concentrations, and fluxes element by element

through the array. Sometimes, different types of membrane elements (e.g., high-rejection in early stages, high-flux or low-fouling in later stages) might be used within an array to optimize performance.

5.4 Enhancing Cross-Flow and Recovery: Concentrate Recycle

Sometimes, especially in smaller systems or when trying to push recovery limits, maintaining adequate cross-flow velocity in the first stage can be challenging with a simple staged array. Concentrate Recycle offers a solution.

5.4.1 The Concept: Sending Concentrate Back

A portion of the final concentrate stream exiting the last stage is routed back (using a recycle pump or relying on existing pressure, though a pump is more common for control) and blended with the incoming raw feed water before it enters the first stage.

5.4.2 Why Recycle? Pros and Cons

Advantages:

- **Increased Cross-Flow:** By adding recycle flow to the raw feed, the total flow entering the first stage is increased, boosting the cross-flow velocity within the elements. This helps control CP and fouling, especially in the critical lead elements.
- **Higher Recovery Potential (Indirectly)**: While recycle itself doesn't fundamentally change the maximum thermodynamic recovery limit set by scaling, the improved hydrodynamics (higher cross-flow) might allow the system to operate slightly closer to that limit more reliably or with less fouling.
- **System Size Optimization:** In some small system designs, recycle might allow for a configuration with fewer pressure vessels than would otherwise be needed to maintain flow.

Disadvantages:

- **Increased Feed Salinity to Stage 1**: The blended feed entering Stage 1 now has a higher TDS concentration than the raw feed water, because it contains recycled concentrate. This increases the osmotic pressure in the first stage, slightly reducing the NDP and potentially lowering the flux per element compared to a non-recycle system operating at the same raw feed flow and pressure.
- **Increased Scaling/Fouling Potential in Feed:** The higher concentration of scalants and foulants in the blended feed might increase the stress on the lead elements if not properly managed with pretreatment.
- **Energy Consumption:** If a recycle pump is used, it consumes additional energy.
- **Complexity:** Adds piping, potentially a pump, and control loops.

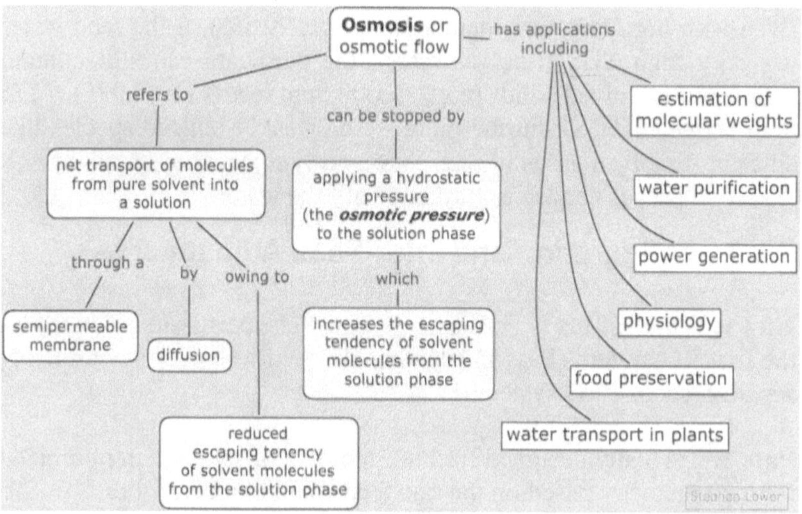

Figure 5.3: Concept Map (Image from Chem1.com)

Use: Recycle is often employed in smaller BWRO systems or sometimes in specific industrial applications. Its use requires careful consideration and calculation using design software to balance the benefits of increased cross-flow against the drawback of increased feed salinity to the membranes.

5.5 Polishing the Product: Double Pass (Two-Pass) RO

For applications demanding extremely high purity water (e.g., boiler feedwater for high-pressure turbines, pharmaceutical water, semiconductor manufacturing), the permeate quality from a single pass-through standard RO membranes might not be sufficient. In these cases, a Double Pass or Two-Pass RO system is used.

5.5.1 When One Pass Isn't Enough

Even with high rejection membranes (e.g., 99.5%), if the feed water has very high TDS (like seawater), the permeate can still contain significant dissolved solids (e.g., 35,000 ppm feed * (1 - 0.995) = 175 ppm permeate TDS). Furthermore, some weakly ionized species like silica or dissolved gases like CO_2 are poorly rejected by standard RO. A second pass is needed to further purify the water.

5.5.2 Configuration and Inter-Pass Adjustments

How it Works: This is Permeate Staging. The permeate produced by the first RO system (Pass 1) becomes the feed water for a completely separate, second RO system (Pass 2).

Pass 1: Typically designed for bulk removal of salts, often optimized for high recovery based on the raw feed water characteristics.

Pass 2: Treats the relatively clean, low-TDS permeate from Pass 1. It typically uses membranes designed for low pressure operation and very high rejection. Because the feed to Pass 2 is already quite pure, its osmotic pressure is low, allowing operation at much lower pressures than Pass 1. Recovery in Pass 2 is usually high (e.g., 90%).

Inter-Pass Adjustments: Sometimes necessary between passes:

- **pH Adjustment:** The permeate from Pass 1 can be acidic due to the passage of dissolved CO_2. Caustic soda (NaOH) may be injected before Pass 2 to increase the pH. This converts dissolved CO_2 to bicarbonate ions (HCO_3^-), which are well rejected by the Pass 2 membranes, resulting in much lower final permeate conductivity and better silica rejection (as silica rejection is better at higher pH).
- **Break Tank (Optional):** A storage tank may be included between passes for operational flexibility or pressure breaking.
- **Pass 2 Feed Pump:** Required to provide the necessary pressure for the Pass 2 membranes.

5.6 Building Capacity and Flexibility: Multiple Trains

For large water treatment plants needing high volumes of purified water, building one single, enormous RO system is usually impractical and inflexible. Instead, the required capacity is achieved by installing Multiple Trains.

5.6.1 Parallel Systems

Definition: A Train is a complete, independent RO system (including pretreatment, pumps, membrane arrays, controls) capable of producing a certain portion of the total required flow. Multiple identical (or sometimes different capacity) trains are installed in parallel, all drawing from a common feed source and delivering permeate to a common storage or distribution system.

Example: A 10 million GPD desalination plant might consist of ten parallel 1 million GPD RO trains

5.6.2 Redundancy and Maintenance Advantages

- **Operational Flexibility:** If the total water demand is less than full capacity, some trains can be shut down to save energy.

- **Redundancy (Reliability):** If one train needs maintenance (e.g., membrane cleaning, pump repair), it can be taken offline without shutting down the entire plant. The other trains continue operating, ensuring a continuous water supply (though possibly at reduced total capacity).
- **Easier Maintenance:** Working on a smaller, self-contained train is generally easier and safer than working on components within a massive, single system.
- **Scalability:** Plants can be built with fewer trains initially and expanded by adding more trains later as demand grows.

Figure 5.4: Large-scale RO plants use multiple membrane cartridges. (Image from Chem1.com)

5.7 Chapter Summary and Key Takeaways

- RO systems require arranging multiple membrane modules within high-pressure vessels.
- Concentrate Staging (arranging vessels in series, e.g., 4:2:1 array) is essential to manage decreasing flow and increasing concentration, maintain cross-flow velocity, and optimize recovery.
- Concentrate Recycle blends reject back with the feed to increase first-stage cross-flow but increases feed salinity to the membranes.
- Double Pass (Two-Pass) RO uses the permeate from Pass 1 as feed for Pass 2 to achieve ultra-high purity, often with inter-pass pH adjustment.
- Large plants use Multiple Trains (independent parallel RO systems) for capacity, operational flexibility, redundancy, and easier maintenance.
- These flow patterns are fundamental design choices determined by feed water quality, desired product quality, recovery targets, and economic considerations, usually optimized using specialized software.

5.8 References

1. AWWA Manual M46: Reverse Osmosis and Nanofiltration (2007). American Water Works Association. (Chapter on System Design covering arrays, staging, and passes).
2. DuPont Water Solutions. (Technical Manuals and Design Guidelines). e.g., WAVE Software Manual, FilmTec™ Membranes Technical Manual. (Provide details on array configurations and design principles).
3. Hydranautics (A Nitto Group Company). (Technical Service Bulletins and Software Manuals). e.g., IMSDesign Software Help. (Illustrate various flow patterns and their implications).
4. Voutchkov, N. (2018). Desalination Engineering: Planning and Design. McGraw-Hill Education. (Covers large-scale plant design, including staging and train configurations).

5. Glueckstern, P., & Priel, M. (1998). Optimization of RO systems design. Desalination, 119(1-3), 121-128. (Example of research paper discussing design optimization including staging).
6. Wilf, M., & Klinko, K. (1998). Performance of commercial seawater membranes. Desalination, 117(1-3), 279-295. (Discusses practical aspects of seawater RO design, implicitly involving staging).

Chapter 6: Reverse Osmosis Skids - The Complete Assembly

6.1 Introduction: From Membranes to Machine

We've established that Reverse Osmosis relies on semi-permeable membranes packaged into modules (elements) and arranged in specific flow patterns (arrays/stages) within pressure vessels. But these core components cannot function in isolation. They need pumps to provide pressure, pipes to direct flow, filters for protection, instruments to monitor conditions, and controls to manage the entire process safely and efficiently.

In most industrial and commercial applications, these components are pre-assembled onto a structural metal frame called a skid. This skid-mounted design offers several advantages:

- **Factory Assembly & Testing:** Allows for controlled assembly and often pre-testing at the manufacturer's facility, reducing on-site installation time and potential errors.
- **Compact Footprint:** Integrates components efficiently, minimizing space requirements.
- **Transportability:** Facilitates easier shipping and placement at the final installation site.
- **Integrated Design:** Ensures components are properly matched and interconnected.

This chapter dissects the typical RO skid, examining each critical component beyond the membranes themselves. We'll explore their function, common types, selection criteria, and operational significance, providing practical insights gleaned from decades of seeing these systems in action. Whether it's a small under-the-counter unit (which is essentially a miniaturized, simplified skid) or a large industrial train, the fundamental building blocks are remarkably similar.

6.2 The Final Guard: Cartridge Filters

Just before the feed water enters the expensive high-pressure pump and the delicate RO membranes, it typically passes through a Cartridge Filter Housing. This is not the primary means of removing suspended solids – that's the job of the upstream pretreatment (media filters, UF, etc.). Instead, the cartridge filter acts as a crucial polishing filter or safety filter.

- **Purpose:** To capture any fine particulate matter that may have slipped through the main pretreatment or originated downstream (e.g., fines from media filters during backwash upsets, fragments from ion exchange resin, small rust particles from piping). Its primary role is to protect the high-pressure pump from abrasion and prevent premature plugging of the lead RO membrane elements by stray particles.
- **Mechanism:** Operates in dead-end filtration mode (Chapter 2). Water flows from the outside of the cylindrical filter cartridges to the inside core.
- **Filter Cartridges:** Typically made of melt-blown or string-wound polypropylene fibers.
    - **Rating:** Usually rated 5-micron (μm) nominal. "Nominal" means it removes most (perhaps 80-90%) of particles larger than that size. Sometimes, 1-micron absolute filters (removing >99.9% of particles larger than the rating) are used for extra protection, but they tend to clog faster and have higher pressure drop.
- **Housing:** Usually a vertical vessel made of stainless steel (common) or FRP, designed to hold multiple cartridges. Must be rated for the pre-pump feed pressure.
- **Operation & Maintenance:**
    - **Monitoring:** The key parameter is the differential pressure (ΔP) across the filter housing (pressure difference between inlet and outlet gauges/transmitters).
    - **Replacement:** Cartridges must be replaced when the ΔP reaches a predetermined limit, typically 10-15 psid (psi differential) above the clean ΔP, or based on a regular time interval if ΔP doesn't increase significantly. Allowing ΔP

to get too high can cause the cartridges to collapse or unload previously trapped particles.

- **Practical Insight:** The frequency of cartridge filter replacement is an excellent indicator of upstream pretreatment performance. If cartridges need changing daily or even multiple times a day, it's a strong sign that the pretreatment system is failing or inadequate, and the root cause needs investigation immediately. Simply changing cartridges more often is a costly band-aid that puts the RO system at risk. A well-performing pretreatment system might allow cartridges to last for weeks or even months.

6.3 The Powerhouse: Reverse Osmosis Feed Pumps

This is the heart of the RO system, providing the hydraulic energy required to overcome the natural osmotic pressure of the feed water and the pressure losses through the system, driving the water molecules through the membrane.

- **Purpose:** To increase the feed water pressure from the supply pressure (after pretreatment and cartridge filtration) to the high pressure required for the RO membranes (ranging from ~100 psi for low-TDS water to >1000 psi for seawater).
- **Types:**
  - **Multi-Stage Centrifugal Pumps:** The workhorse for most commercial and industrial RO systems. Multiple impellers and diffusers are arranged in series on a single shaft, with each stage adding a portion of the total pressure boost. They offer smooth, continuous flow and relatively high efficiency over a range of operating points. Typically horizontal or vertical configurations.
  - **Positive Displacement (PD) Pumps:** Such as reciprocating plunger or piston pumps. These deliver a fixed volume of water per stroke or rotation, making flow relatively independent of pressure. They can achieve very high pressures efficiently and are sometimes used for very high-pressure SWRO or smaller systems. However, they produce pulsating flow (requiring pulsation dampeners), can be noisier, require more maintenance, and generally

have a higher capital cost for larger flows compared to centrifugals.

- **Key Selection Criteria:**
    - **Required Pressure (Head):** Determined by feed water TDS (osmotic pressure), temperature, desired recovery rate, membrane type, and system pressure losses. Must be calculated accurately by design software.
    - **Required Flow Rate:** Determined by the target permeate production rate and the designed recovery rate (Feed Flow = Permeate Flow / Recovery Rate).
    - **Efficiency:** Pumps are major energy consumers (often the largest in an RO system). Selecting a pump that operates near its Best Efficiency Point (BEP) for the typical operating conditions is crucial for minimizing energy costs.
    - **Materials of Construction:** Must withstand the operating pressure and be resistant to corrosion from the feed water, especially chlorides. Common materials include 316L Stainless Steel, Duplex Stainless Steel, or Super Duplex Stainless Steel for seawater or high-chloride brackish water.
    - **Net Positive Suction Head (NPSH):** The pump needs sufficient inlet pressure (NPSH Available > NPSH Required) to avoid cavitation (formation and collapse of vapor bubbles), which can severely damage the pump impellers. This involves considering feed tank elevation, pipe friction losses, and water temperature.
- **Control:** Modern systems almost always utilize Variable Frequency Drives (VFDs) to control the speed of the centrifugal feed pump motor. This allows for:
    - **Soft Starting:** Reduces electrical inrush current and mechanical stress.
    - **Precise Flow/Pressure Control:** Allows the system to adjust output based on demand or maintain a constant flux despite changing feed conditions (like temperature), often more efficiently than using a throttling valve.
- **Practical Insight:** The feed pump is often the single most expensive component on the skid and the largest energy user. Proper selection, operation near BEP, and preventative

maintenance are critical for reliable and cost-effective RO operation. Understanding the pump performance curve (Head vs. Flow, Efficiency vs. Flow) is essential for operators.

6.4 Containing the Pressure: Pressure Vessels Revisited

We discussed pressure vessels in Chapter 5 as the containers for membrane elements. On the skid, they represent a significant part of the high-pressure infrastructure.

- **Recap:** Function is to house 1-8 elements in series under pressure. FRP is the most common material due to corrosion resistance and strength. Standard diameters (e.g., 4-inch, 8-inch) and lengths (for 40-inch long elements) are typical.
- **Skid Integration:** Vessels are mounted horizontally (most common) or sometimes vertically onto the skid frame. They are connected via the feed/concentrate manifolding.
- **Key Features on Skid:**
  - **End Caps/Heads:** Secured by locking mechanisms (e.g., retaining rings, couplers). Must be rated for the maximum system pressure. Provide ports for feed/concentrate and permeate connections. O-rings ensure sealing.
  - **Saddle Supports:** Secure the vessel to the skid frame.
  - **Safety:** Vessels operating above certain pressures/volumes must often comply with codes like the ASME Boiler and Pressure Vessel Code, ensuring safe design and construction. Burst discs or pressure relief valves may be incorporated in the system design.
- **Practical Insight:** Proper alignment during installation is crucial to avoid stress on the vessels and end caps. Regular inspection of end cap seals and securing mechanisms is important for safety. Loading and unloading elements requires care to avoid damaging the elements or vessel interiors.

6.5 The Plumbing Network: Manifolding and Materials of Construction

The **manifolding** is the system of pipes and fittings on the skid that distributes feed water to the pressure vessels and collects the permeate and concentrate streams. Choosing the right materials is critical for longevity and safety.

- **Purpose:** To transport water between components (pump, filters, vessels, instruments) safely and efficiently at various pressures and flow rates.
- **Material Selection Criteria:**
  - **Pressure Rating**: Must withstand the maximum operating pressure (with a safety factor).
  - **Corrosion Resistance:** Crucial, especially for feed and concentrate streams containing salts, chlorides, and potentially treatment chemicals. Compatibility with cleaning chemicals must also be considered.
  - **Temperature Rating:** Must handle the operating temperature range.
  - **Cost:** Varies significantly between materials.
  - **Ease of Fabrication/Installation:** Affects labor costs.
- **Common Materials:**
  - **Low Pressure Piping (Permeate, Drain, Chemical Lines < ~100 psi):**
    - **PVC (Polyvinyl Chloride - Schedule 80):** Inexpensive, easy to join (solvent cement), good chemical resistance, but limited temperature/pressure rating and can become brittle over time.
    - **CPVC (Chlorinated PVC):** Higher temperature resistance than PVC.
    - **PP (Polypropylene), PVDF (Polyvinylidene Fluoride):** Excellent chemical resistance, often used for chemical dosing lines. Joined by fusion welding.
  - **High Pressure Piping (Feed, Concentrate, Interstage > ~100 psi):**

    - **316L Stainless Steel:** Common choice for brackish water RO due to good corrosion resistance (low carbon 'L' grade preferred for welding). Typically Schedule 10 or 40 pipe. Joined by welding (TIG preferred) or specialized mechanical couplings (e.g., Victaulic). Susceptible to pitting/crevice corrosion in high chloride environments (> ~1000 ppm Cl^-).
    - **Duplex and Super Duplex Stainless Steels (e.g., 2205, 2507):** Significantly higher strength and much better resistance to chloride pitting and stress corrosion cracking than 316L. Standard for seawater RO and high-chloride brackish water. More expensive and requires more controlled welding procedures.
    - **Specialized Alloys (e.g., Hastelloy, Monel):** Used in highly corrosive industrial applications, very expensive.
    - **FRP or Lined Steel Pipe:** Sometimes used, especially for large diameter headers.
- **Design Considerations:**
    - Proper pipe sizing to minimize pressure drop (friction losses).
    - Use of long-radius bends instead of sharp elbows where possible.
    - Inclusion of isolation valves (ball valves, butterfly valves) for maintenance on vessels, pumps, instruments.
    - Sample ports for water quality testing.
    - Appropriate supports to prevent vibration and stress.
    - Avoiding dissimilar metal contact where possible to prevent galvanic corrosion.
- **Practical Insight:** Leaks, particularly at joints and valve connections, are common issues. Proper installation technique (welding, solvent cementing, gasket seating) is vital. Material selection errors (e.g., using 316SS in high-chloride water) can lead to rapid and costly failures. Manifold design should allow easy access for maintenance.

6.6 The System's Senses: Instrumentation

To understand and control what's happening within the RO system, accurate measurements are essential. Instrumentation provides the real-time data needed for operation, monitoring, and troubleshooting.

Purpose: To measure key process variables like pressure, flow, conductivity, temperature, pH, etc.

Common Instruments and Locations: (See Table 6.1)

- **Pressure Gauges/Transmitters:**
  - Cartridge Filter Inlet/Outlet (for ΔP)
  - Pump Suction/Discharge
  - RO Feed Header (before vessels)
  - Interstage Headers (between stages in staged arrays)
  - Concentrate Header (after last stage)
  - Permeate Header
- **Flow Meters:**
  - Feed Flow (after pump)
  - Permeate Flow (total)
  - Concentrate Flow (total)
  - Recycle Flow (if applicable)
- **Conductivity/TDS Sensors:**
  - Feed Water (often after pretreatment)
  - Permeate (CRITICAL for quality monitoring)
  - Interstage/Concentrate (optional, for diagnostics)
- **Temperature Sensors:**
  - Feed Water (essential for performance normalization)
- **pH Sensors:**
  - Feed Water (important if pH adjustment is used or feed pH varies)
  - Permeate (optional, can indicate CO_2 passage)
- **ORP (Oxidation-Reduction Potential) Sensors:**
  - Feed Water (typically placed just before membranes, after any dechlorination step, to ensure chlorine/oxidant removal is complete - target < 150-200 mV usually).
- **Level Switches/Transmitters:**

  - Feed Water Tank
  - Permeate Storage Tank
  - Cleaning Tank (CIP)
- **SDI Test Assembly:** Manual kit or automated unit connected to the RO feed line (after cartridge filter).

Types of Instruments: Range from simple visual gauges and rotameters to sophisticated electronic transmitters (4-20mA output) connected to the control system. Common flow meter types include turbine, magnetic flow (magmeter - good for conductive water, no moving parts), paddlewheel, ultrasonic, and variable area (rotameter).

Practical Insight: Instrumentation is only useful if it's accurate! Regular calibration is non-negotiable. Location matters – sensors must be installed correctly to get representative readings (e.g., sufficient straight pipe run for flow meters). Redundant sensors for critical parameters like permeate conductivity can provide backup and cross-checking. Investing in reliable instrumentation pays off in better control and faster troubleshooting.

Table 6.1: Common RO Skid Instrumentation

Parameter	Typical Location(s)	Purpose	Common Instrument Type(s)
Pressure	Filter In/Out, Pump Suct/Disc, Feed, Interstage, Conc, Perm	Monitor driving force, pressure drops (ΔP), system limits	Gauge (Visual), Transmitter (4-20mA)
Flow	Feed, Permeate, Concentrate, Recycle	Measure production, calculate recovery, monitor pump performance	Turbine, Magnetic, Paddlewheel, Ultrasonic, Rotameter
Conductivity/ TDS	Feed, Permeate	Monitor feed quality, CRITICAL for permeate quality (rejection)	Probe/Transmitter (Temp Compensated)

Temperature	Feed	Essential for performance normalization (flux, rejection calculations)	RTD, Thermocouple / Transmitter
pH	Feed, Permeate (Optional)	Monitor feed stability, optimize cleaning, check CO_2 passage	Probe/Transmitter
ORP	Feed (after dechlorination, before membranes)	Verify absence of harmful oxidants (e.g., chlorine)	Probe/Transmitter
Level	Tanks (Feed, Permeate, CIP)	Control pump operation, prevent overflow/running dry	Float Switch, Ultrasonic, Radar, Pressure
SDI	Feed (after cartridge filter)	Assess particulate fouling potential (manual test or automated unit)	Filter Pad Assembly

6.7 The System's Brain: Controls and Automation

The control system integrates the signals from the instrumentation and operates the pumps, valves, and alarms to run the RO system automatically, safely, and according to desired setpoints.

- **Purpose:** To automate system operation (start/stop, flushing), maintain target performance (pressure, flow), protect equipment (shutdowns on alarm), and provide an interface for the operator.
- **Levels of Automation:**

  - **Manual:** Operator manually starts/stops pumps, adjusts valves based on gauge readings (rare except for very small or old systems).
  - **Semi-Automatic:** Basic start/stop logic, some automated valves, basic alarms.
  - **Fully Automatic:** PLC/DCS based control, automated sequences, PID control loops, comprehensive alarming, data logging, remote monitoring capabilities. This is the standard for modern industrial RO.
- **Control Hardware:**
  - **PLC (Programmable Logic Controller):** The core processing unit. Reads inputs from sensors, executes the programmed logic, and sends outputs to control devices (pumps, valves, indicators). Robust, industry-standard controllers (e.g., Allen-Bradley, Siemens, Modicon).
  - **HMI (Human-Machine Interface):** Typically a touchscreen display mounted on the control panel. Allows operators to view system status, monitor variables, change setpoints, acknowledge alarms, and manually control components. Provides graphical representation of the process (P&ID view).
  - **Control Panel:** Enclosure housing the PLC, HMI, power supplies, relays, motor starters/VFDs, circuit breakers, terminal strips.
- **Key Control Functions/Logic:**
  - **Start-up Sequence:** Timed valve openings, pump soft start (VFD), gradual pressure ramping. Often includes a low-pressure flush.
  - **Normal Operation:** Maintain target feed pressure or permeate flow using VFD control or modulating valves. Monitor key parameters against alarm limits.
  - **Shutdown Sequence:** Gradual ramp down, timed valve closures. Crucially includes a low-pressure permeate flush cycle where permeate water is used to flush feed water out of the membranes before extended shutdown, minimizing scaling and biofouling potential during standstill.
  - **Alarm Handling:** Detects abnormal conditions (e.g., high/low pressure, low flow, high permeate conductivity, high filter ΔP, motor overload, tank level issues, leak

detection). Triggers audible/visual alarms and initiates appropriate shutdown or protective actions.
    - **Valve Sequencing:** Automatic operation of feed, concentrate, permeate, flush, and sometimes cleaning valves based on operating mode.
- **Practical Insight:** A well-programmed control system is crucial for protecting the membranes and equipment, ensuring consistent water quality, and minimizing operator intervention. The logic should be robust, accounting for various operating scenarios and potential faults. Clear alarm descriptions and troubleshooting guidance within the HMI are invaluable for operators.

6.8 Tracking Performance: Data Acquisition and Management

Collecting and analyzing operating data is essential for understanding how the RO system is performing over time, identifying trends (like fouling), optimizing performance, scheduling maintenance, and troubleshooting problems.

- **Purpose:** To systematically record, store, display, and analyze key operating parameters.
- **Methods:**
    - **Manual Log Sheets:** Basic method, operator records readings periodically. Prone to errors, difficult to analyze trends.
    - **Local Data Logging (PLC/HMI):** PLC stores data internally or on a memory card. HMI can display trends over limited periods.
    - **SCADA (Supervisory Control and Data Acquisition) Systems:** Centralized computer system that communicates with the PLC(s), collects data, displays real-time information graphically, archives data in a historian database, and manages alarms. Common for larger plants.
    - **Cloud-Based Platforms:** Increasingly popular. Data is sent from the PLC/SCADA system to secure cloud servers

for long-term storage, remote access, advanced analytics, reporting, and predictive maintenance capabilities.

- **Key Data Points to Log:** All the major instrument readings (flows, pressures, conductivity, temperature, ORP, pH), calculated values (flux, rejection, recovery), valve positions, pump status/speed, alarm history, run hours. Sampling frequency depends on the parameter (e.g., pressures/flows every minute, quality parameters maybe every 5-15 minutes).
- **Performance Normalization:** This is absolutely critical! Raw permeate flow and salt rejection values change significantly with feed water temperature, pressure, and salinity. Simply plotting raw flow decline might trigger unnecessary cleanings in winter (due to cold water) or mask actual fouling in summer. Normalization mathematically adjusts the current operating data (flux, rejection, pressure drop) back to a standard set of reference conditions (usually initial startup conditions or standard temperature/pressure like 25°C/77°F). Plotting normalized performance over time reveals the true impact of fouling or membrane degradation, independent of changing operating conditions. RO design software or specialized data analysis tools perform these calculations.
- **Practical Insight:** Don't just collect data, use it. Regularly review trend plots of normalized flux, normalized salt passage (or rejection), and differential pressures. A gradual decline in normalized flux or increase in normalized salt passage/ΔP is the classic signature of fouling and indicates cleaning is needed before performance degrades drastically. Good data management provides justification for maintenance actions and helps diagnose the type of fouling occurring.

6.9 The Framework: The Reverse Osmosis Skid Itself

This is the structural backbone that holds everything together.

- **Construction:** Typically fabricated from welded structural steel (e.g., carbon steel I-beams, channels, square tubing).

Sometimes stainless steel is used for highly corrosive environments or sanitary applications.

- **Finish:** Carbon steel skids are usually sandblasted and painted with industrial coatings (e.g., epoxy primer, polyurethane topcoat) for corrosion protection.
- **Design:**
  - **Structural Integrity:** Must safely support the weight of all components, including water-filled vessels and pipes, and withstand operating vibrations (especially from pumps).
  - **Layout:** Designed for logical flow, minimizing piping runs and pressure drops. Crucially, must provide adequate access for operation and maintenance (e.g., space to remove pump motors, load/unload membrane elements, access instruments for calibration, open electrical panels).
  - **Footprint:** Optimized for plant space requirements.
  - **Transportability:** Include lifting lugs and designed to fit standard shipping containers or trucks if necessary.
  - **Containment:** Often includes a raised lip or drip pan to contain minor leaks and spills, with a drain connection.
  - **Vibration Damping:** Pads may be used under pumps to minimize vibration transmission.

6.10 Essential Partners: Auxiliary Equipment (CIP, Chemical Dosing)

While not always mounted directly on the main RO skid (especially for larger systems), certain auxiliary systems are indispensable for operation.

- **Cleaning Skid (CIP - Clean-In-Place):** RO membranes require periodic chemical cleaning to remove accumulated foulants and scale. A dedicated CIP skid provides the necessary equipment:
  - **CIP Tank:** To mix cleaning chemicals (acids, bases, detergents, biocides) with permeate water. Often includes a heater to warm the solution (cleaning is more effective

at higher temperatures, typically up to the membrane limit, e.g., 35-45°C).

- **CIP Pump:** Low-pressure, high-flow pump to circulate the cleaning solution through the RO membranes (bypassing the main feed pump). Material compatibility with cleaning chemicals is key.
- **Cartridge Filter:** To remove any dislodged solids during cleaning before they return to the membranes.
- **Instrumentation:** Temperature gauge/control, level indicator, sometimes pH/conductivity monitoring.
- **Connections:** Hoses and quick-connect fittings to easily connect the CIP skid to designated ports on the RO skid manifolding.

- **Chemical Dosing Skids:** Most RO systems require continuous injection of specific chemicals into the feed water before the membranes:
  - **Antiscalant Dosing:** Prevents mineral scaling ($CaCO_3$, $CaSO_4$, etc.). Consists of a chemical storage tank (day tank), a metering pump (positive displacement type like diaphragm or peristaltic for accurate low-flow dosing), calibration column (to verify dose rate), pressure relief valve, and injection quill into the feed line.
  - **Dechlorination Dosing (e.g., SBS):** If feed water is chlorinated, sodium bisulfite (SBS) is dosed to neutralize chlorine before it reaches TFC membranes. Similar setup to antiscalant skid. Requires ORP monitoring downstream to confirm complete dechlorination.
  - **pH Adjustment Dosing (Acid/Caustic):** Sometimes used to control $CaCO_3$ scale (acid dosing) or optimize rejection/cleaning (caustic dosing). Requires careful pH monitoring and control.
  - **Biocide Dosing:** For intermittent or continuous control of biological growth, though often managed through pretreatment or cleaning rather than continuous dosing to the RO feed.

6.11 Enhancing Operation: Other Design Considerations

Beyond the basic components, certain design features can significantly improve the ability to monitor, troubleshoot, and maintain the RO system, especially for larger or multi-stage units.

6.11.1 Pinpointing Problems: Access to Profile and Probe RO Membranes

- **Profiling:** Involves installing sample ports on the permeate outlet of each individual pressure vessel. By collecting and analyzing the conductivity (or specific ions) from each vessel's permeate, operators can identify which vessel(s) might contain a leaking O-ring, a damaged membrane element, or localized fouling causing poor rejection.
- **Probing:** A more advanced technique, requiring specialized end caps with probe ports. Allows insertion of a thin tube down the central permeate tube of the elements within a vessel (while operating or offline) to collect permeate samples from specific points along the vessel length. This can pinpoint a problem to an individual element or even a specific location within an element.
- **Value:** Invaluable for troubleshooting large arrays. Without profiling/probing, finding a single leaking element among hundreds might require systematically dismantling vessels – a time-consuming and costly process.

6.11.2 Deeper Insight: Interstage Performance Monitoring

- **Concept:** Installing pressure transmitters and sometimes flow or conductivity sensors on the manifolding between the stages of a multi-stage array (e.g., between Stage 1 concentrate outlet and Stage 2 feed inlet).
- **Value:** Allows calculation of performance metrics (ΔP, recovery, rejection) for each stage independently, rather than

just the overall system. This helps identify which stage is experiencing the most significant fouling (e.g., if Stage 1 ΔP increases rapidly) or if performance issues are localized.

6.11.3 Targeted Cleaning: Stage-by-Stage Cleaning Provisions

- **Concept:** Incorporating isolation valves and separate CIP supply/return connections in the manifolding to allow individual stages (or sometimes even individual vessels) to be chemically cleaned independently while the rest of the system might remain offline or (in some advanced designs) even operating.
- **Value:** Useful if fouling tends to occur preferentially in one stage (e.g., biofouling often worst in the tail stage). Allows targeted use of cleaning chemicals and potentially reduces cleaning downtime for the entire system. Requires more complex and expensive piping/valving.

6.12 Chapter Summary and Key Takeaways

- An RO Skid integrates membranes, pressure vessels, pumps, filters, piping, instruments, and controls onto a structural frame to create a functional unit.
- Cartridge filters provide essential protection for the high-pressure pump and lead membranes from stray particulates. Monitor ΔP closely.
- The RO feed pump (multi-stage centrifugal or PD) provides the necessary pressure; efficiency and material selection are critical. VFDs are standard for control.
- Pressure vessels (usually FRP) house the membrane elements safely under pressure.
- Manifolding requires careful material selection (based on pressure, corrosion resistance – e.g., 316L SS, Duplex SS) and design to ensure reliable flow distribution.
- Instrumentation (pressure, flow, conductivity, temperature, pH, ORP, level) provides the "senses" needed for monitoring and control. Calibration is key.

- The Control System (PLC/HMI) automates operation, ensures safety through alarms/shutdowns, and executes sequences like permeate flushing.
- Data Acquisition and Performance Normalization are vital for tracking true membrane performance (fouling) over time, independent of operating condition changes.
- Auxiliary skids for CIP (cleaning) and chemical dosing (antiscalant, dechlorination) are essential partners to the main RO skid.
- Advanced design features like profiling/probing, interstage monitoring, and stage-by-stage cleaning provisions greatly enhance troubleshooting and maintenance capabilities, particularly on larger systems.

6.13 References

1. AWWA Manual M46: Reverse Osmosis and Nanofiltration (2007). American Water Works Association. (Chapters on System Components, Instrumentation, and Controls).
2. DuPont Water Solutions, Hydranautics, Toray, LG Chem, etc. (Manufacturer Technical and Engineering Manuals): Provide detailed information on skid components associated with their membrane products, including pressure vessel specifications, instrumentation recommendations, and CIP procedures.
3. Pump Handbook (e.g., by Karassik, Messina, Cooper, Heald). (Comprehensive resource on pump types, selection, and operation).
4. Instrument Engineers' Handbook (e.g., by Béla Lipták). (Detailed information on various types of process instrumentation).
5. ASME Boiler and Pressure Vessel Code. (Relevant sections for pressure vessel design and safety).
6. Voutchkov, N. (2018). Desalination Engineering: Planning and Design. McGraw-Hill Education. (Covers practical aspects of plant components and skid design).

Chapter 7: RO Design and Design Software - Predicting Performance

7.1 Introduction: From Theory to Blueprint

We've journeyed through the fundamental principles, dissected the membranes and modules, examined flow patterns, and surveyed the components on an RO skid. Now, we arrive at the synthesis: the design phase. How do we take a specific water source, a desired product quality, and a required flow rate, and translate them into a reliable, efficient, and cost-effective Reverse Osmosis system specification?

While modern design heavily relies on sophisticated computer software, a solid understanding of the underlying calculations is paramount. It allows engineers and experienced operators to perform "sanity checks" on software outputs, understand why a system is designed a certain way, and make informed decisions during operation and troubleshooting. Design is often an iterative process – initial calculations or simulations reveal potential issues (like scaling or excessive flux), requiring adjustments to parameters like recovery rate, membrane choice, or pretreatment strategy until an optimal and safe design is achieved.

This chapter first revisits the key calculations that form the basis of RO design and then dives into the practical application of membrane manufacturer design software, covering the essential inputs, parameters, and critical outputs that must be carefully interpreted.

7.2 The Building Blocks: Key Design Calculations

Before the advent of readily available software, engineers relied on manual calculations, often using simplified assumptions or graphical methods. While tedious for complex systems, understanding these basic calculations provides valuable insight.

7.2.1 Osmotic Pressure (π): The Back-Pressure Barrier

Concept: As covered in Chapter 2, this is the pressure generated by the dissolved salts, opposing the flow of water across the membrane. It's the primary resistance the feed pump must overcome.

Calculation: Estimated using the Van't Hoff equation (π = iCRT) or rules-of-thumb (approx. 1 psi per 100 mg/L TDS). Crucially, it needs to be calculated not just for the initial feed water (πf) but also for the concentrated water at various points along the membrane path, especially the final concentrate (πc).

Significance: Directly impacts the required feed pressure. Higher TDS = higher π = higher required feed pressure = higher energy consumption.

7.2.2 Net Driving Pressure (NDP): The Real Push

Concept: The actual pressure difference available to drive water through the membrane (P feed - P permeate - $\Delta\pi$).

Calculation: Needs to account for the average osmotic pressure difference across the membrane surface, considering concentration polarization (CP). Software models this more accurately.

Significance: Water flux is directly proportional to NDP (Jw = A * NDP). NDP decreases along the flow path as feed pressure drops (friction) and osmotic pressure increases (concentration). The design must ensure sufficient NDP exists even in the last elements.

7.2.3 Membrane Area & Flux: Sizing the System

Concept: Flux (Jw) is the flow rate per unit area (GFD or LMH). To achieve a target permeate flow rate (Qp), a specific total membrane area (Area) is required.

Calculation: Area = Qp / J w,avg. The challenge lies in determining the appropriate average design flux (J w,avg), which depends on membrane type, water quality (fouling potential), temperature, and balancing capital cost (less area = fewer elements = lower cost) against operational reliability (lower flux = less fouling = longer runs between cleanings).

Significance: Determines the number of membrane elements and pressure vessels needed, directly impacting system footprint and capital cost. Software calculates flux element-by-element, ensuring local flux limits aren't exceeded.

7.2.4 Recovery, Beta, and Salt Balance: Mass Matters

Concept: Recovery (R = Qp / Qf) dictates how much feed becomes product. The concentration factor ($\beta \approx 1 / (1 - R)$) quantifies how much the salts concentrate in the reject stream (Qc). A simple salt mass balance (Salt In = Salt Out) must hold: Qf * Cf = Qp * Cp + Qc * Cc.

Calculation: These are algebraically linked. Setting a target recovery allows calculation of Qf, Qc, and β. Using the membrane's expected salt rejection (Rej = 1 - Cp/C f,avg), the permeate concentration (Cp) and concentrate concentration (Cc) can be estimated.

Significance: Recovery is a primary driver of concentrate concentration (β), which in turn dictates scaling potential. The salt balance allows prediction of permeate quality.

7.2.5 Scaling Indices (LSI, S&DSI, Silica Saturation): The Critical Limits

Concept: As water concentrates, sparingly soluble salts can precipitate (scale). Indices predict this tendency.

Calculation: Requires the concentrate water composition (calculated by multiplying feed concentrations of key ions like Ca^{2+}, HCO_3^-, SO_4^{2-}, SiO_2 by the concentration factor β) and plugging these values, along with concentrate pH and temperature, into the relevant index formulas:

- **Langelier Saturation Index (LSI):** For $CaCO_3$ scale. (LSI = pHc - pHs)

- **Stiff & Davis Stability Index (S&DSI):** Better for higher TDS waters predicting $CaCO_3$, sometimes adapted for $CaSO_4$.
- **Calcium Sulfate ($CaSO_4$) Solubility Product:** Calculate ion product $[Ca^{2+}]*[SO_4^{2-}]$ in the concentrate and compare to the solubility product constant (Ksp) at that temperature and ionic strength.
- **Silica (SiO_2) Saturation:** Compare calculated concentrate silica concentration to its temperature-dependent solubility limit (amorphous silica solubility is complex and pH-dependent above ~pH 8).

Significance: These calculations determine the maximum allowable recovery rate for a given feed water and pretreatment strategy. Exceeding these limits invites rapid scaling and membrane damage. If the target recovery leads to unacceptable scaling predictions, the options are: lower the recovery, improve pretreatment (add/increase antiscalant dose), or adjust pH.

7.2.6 Pressure Drop (ΔP): Hydrodynamic Checks

Concept: Water flowing through the feed spacers, fittings, and manifolds encounters friction, causing pressure loss (ΔP).

Calculation: Estimated using fluid dynamics equations (e.g., Darcy-Weisbach) considering channel geometry (spacer thickness), flow velocity, and fluid properties (viscosity). Software models incorporate empirical data for specific element types.

Significance: Total ΔP contributes to the required feed pressure. High ΔP consumes energy and can indicate fouling (blocked channels). ΔP per element and per stage are monitored. A certain minimum ΔP is actually needed to ensure reasonably uniform flow distribution across the parallel elements within a stage.

7.2.6 Cross-flow Velocity:

Concept: The speed at which feed water sweeps across the membrane surface, crucial for controlling concentration polarization and fouling.

Calculation: Feed flow rate through a vessel divided by the cross-sectional area of the feed channels. Staging helps maintain adequate velocity as overall flow decreases.

Significance: Minimum cross-flow velocities are recommended by manufacturers, especially for the last stage where concentration is highest. Low velocity leads to rapid fouling.

7.2.7 The Iterative Dance

As highlighted, these factors are interconnected. Adjusting recovery changes β, impacting scaling and osmotic pressure. Changing pressure affects flux, which influences CP and potentially fouling. Choosing a different membrane changes flux and rejection characteristics. Therefore, manual design involves iterative calculations and adjustments until all parameters fall within acceptable limits – a process perfectly suited for computer simulation.

7.3 The Essential Tool: Use of Design Software

While understanding the calculations is vital, modern RO system design relies almost exclusively on specialized computer software, typically provided free of charge by major membrane manufacturers.

7.3.1 Why Software is Indispensable

Speed and Complexity: Performs the iterative calculations for multi-stage, multi-element arrays almost instantaneously, handling complex interactions that are impractical manually.

Accuracy: Incorporates sophisticated models for osmotic pressure (considering non-ideal solutions), concentration polarization, temperature corrections, and pressure drop specific to the manufacturer's membrane elements.

Proprietary Data: Contains detailed performance characteristics (A and B values, flux limits, etc.) for specific membrane products, which are often proprietary and essential for accurate predictions.

Scenario Analysis: Allows easy comparison of different membrane types, array configurations, operating conditions, and pretreatment strategies ("what-if" scenarios).

Standardized Reporting: Generates comprehensive design reports documenting inputs, assumptions, and predicted performance in a standard format.

7.3.2 Common Software Packages:

Each major membrane manufacturer offers its own sophisticated package:

WAVE (Water Application Value Engine) by DuPont: Successor to ROSA (Reverse Osmosis System Analysis). Comprehensive, widely used.

IMSDesign (Integrated Membrane Solutions Design) by Hydranautics (A Nitto Group Company).

TorayDS2 by Toray.

Q+ Projection Software by LG Chem.

(Note: While interfaces differ, the core principles and required inputs/outputs are generally similar across these platforms.)

7.3.3 The Simulation Process: Inputs, Calculations, Outputs

Using the software involves three main steps:

1. **Defining Inputs:** Telling the software about the feed water and the desired system setup.
2. **Running the Simulation:** The software performs the complex calculations based on its internal models and membrane data.
3. **Interpreting Outputs:** Analyzing the detailed report generated by the software to verify the design's feasibility and safety.

7.4 Step 1: Inputting the Water Analysis - Garbage In, Garbage Out

This is arguably the most critical step. The accuracy of the entire design projection hinges on the quality of the feed water analysis provided. An inaccurate or incomplete analysis will lead to an incorrect design, potentially resulting in underperformance, premature fouling/scaling, or even catastrophic failure.

7.4.1 The Crucial Foundation: Accurate and Complete Data

Accuracy: Use data from a reputable laboratory experienced in water analysis.

Completeness: Ensure all necessary ions for calculating ionic balance and scaling potential are included.

Recency: Water chemistry can change seasonally or due to source variations. Use the most recent analysis possible, representing the typical or worst-case conditions the RO plant will encounter.

Representativeness: Ensure the sample accurately reflects the water that will actually be fed to the RO system after any upstream processes that might alter its chemistry (e.g., softening, clarification).

7.4.2 Key Chemical Parameters Required

The software needs concentrations (usually in mg/L or ppm) of:

- **Major Cations:** Calcium (Ca^{2+}), Magnesium (Mg^{2+}), Sodium (Na^{+}), Potassium (K^{+})
- **Major Anions:** Bicarbonate (HCO_3^-) and/or Alkalinity (as $CaCO_3$), Sulfate (SO_4^{2-}), Chloride (Cl^-). Carbonate (CO_3^{2-}) is important if pH > 8.3.
- **Silica (SiO_2):** Crucial for predicting silica scaling. Specify if measured as reactive or total.
- **pH:** Measured at the sampling temperature.

- **Temperature:** Design temperature or range (see below).
- **Other Potential Scalants/Foulants:** Depending on the source water and application, may also need: Barium (Ba^{2+}), Strontium (Sr^{2+}), Iron (Fe^{2+}/Fe^{3+}), Manganese (Mn^{2+}), Fluoride (F^-), Phosphate (PO_4^{3-}).
- **TDS (Total Dissolved Solids):** Often calculated by the software from the ionic constituents, but providing a measured TDS is a good cross-check for the completeness of the analysis. An ionic balance check (% difference between total cations and anions in meq/L) should ideally be within ±5%.

7.4.3 Temperature: The Dynamic Duo (Flow and Scaling)

Temperature profoundly affects both membrane performance and scaling tendency:

Flux: Water viscosity decreases significantly with increasing temperature, leading to higher flux (more permeate flow). Designs must meet flow requirements at the lowest expected operating temperature.

Scaling: Solubility of some scalants changes with temperature (e.g., $CaCO_3$ solubility decreases as temperature rises, while $CaSO_4$ and SiO_2 solubility generally increases). Scaling potential must be checked at the highest expected operating temperature.

Input: Software typically requires a design temperature (often the average or lowest) and allows checking performance at different temperatures.

7.4.4 Units and Consistency

Double-check that all concentrations are entered in the units expected by the software (usually mg/L=ppm). Ensure alkalinity is specified correctly (as $CaCO_3$ or HCO_3^-).

Practical Insight: If unsure about feed water variability, design based on the worst-case analysis (highest scaling potential, lowest temperature for flow). Consider investing in multiple analyses over time if seasonal variations are suspected. A mistake here can doom the project.

7.5 Step 2: Setting Design Parameters - Defining the Goals

Once the water chemistry is defined, the user specifies the desired system configuration and performance targets.

7.5.1 Membrane Selection:

Choose the manufacturer and the specific membrane element model(s). This involves considering:

- Feed water type (brackish vs. seawater)
- Required rejection (standard vs. high rejection)
- Energy considerations (low energy elements)
- Fouling potential (fouling resistant elements)
- Specific contaminant removal (e.g., low boron elements)
- Cost

7.5.2 System Configuration:

Define the physical layout:

- **Array Design:** Number of stages and number of pressure vessels per stage (e.g., 4:2:1).
- **Elements per Vessel**: Typically 6 or 7 for large systems.
- **Passes:** Single pass or two passes (for high purity).
- **Recycle:** Specify if concentrate recycle is used and the percentage of recycle flow.

7.5.3 Performance Targets:

Permeate Flow Rate: The primary output goal (e.g., GPM, m^3/hr).

Recovery Rate: Enter a target recovery. The software will calculate performance based on this, but it must be checked against scaling limits in the output. Often requires iteration.

7.5.4 Pretreatment Assumptions:

Tell the software how the feed water will be conditioned:

- **pH Adjustment:** If acid is dosed, specify the target feed pH.
- **Antiscalant:** Select the specific antiscalant product from the software's database (if available) or manually input its scaling inhibition limits (e.g., maximum allowable LSI, $CaSO_4$ saturation) based on vendor data. Specify the intended dose rate.

Practical Insight: Start with a reasonable configuration based on experience or preliminary calculations. The software allows easy modification. For example, if the initial recovery target leads to scaling, the designer can easily reduce the recovery, change the antiscalant, or adjust pH in the software and re-run the simulation.

7.6 Step 3: Interpreting Design Outputs - The Moment of Truth

After running the simulation, the software generates a detailed report. This requires careful analysis to ensure the design is viable, safe, and meets specifications.

7.6.1 The Design Report: Key Sections

Reports typically include:

- Input Summary (water analysis, design parameters)
- System Summary (overall flow, recovery, pressure, flux, power)
- Permeate Quality Summary (TDS, individual ion concentrations)
- Scaling Calculation Summary (LSI, S&DSI, Silica, $BaSO_4$, $SrSO_4$ saturation % in the final concentrate)

- Element-by-Element Performance Data (flux, pressure, concentration, rejection for each element in the array)
- Stage-by-Stage Summaries
- Warnings/Flags

7.6.2 Critical Checks: Does it Work? Is it Safe?

Scrutinize these outputs:

1. **Overall Performance:** Does the predicted permeate flow meet the target? Is the required feed pressure reasonable and achievable with standard pumps? Is the estimated power consumption acceptable?
2. **Permeate Quality:** Does the predicted permeate TDS and specific ion concentrations meet the required water quality specifications for the intended use?
3. **Scaling Potential (THE MOST IMPORTANT CHECK):** Are the calculated scaling indices (LSI, $CaSO_4$ %, Silica %, etc.) for the concentrate of the last element within the limits recommended by the membrane manufacturer AND the antiscalant vendor for the specified product and dose? If any limit is exceeded, the design is unsafe and must be revised (lower recovery, change/increase antiscalant, adjust pH). Do not proceed with a design that predicts scaling!
4. **Flux Distribution:** Check the flux of the very first element in Stage 1. Does it exceed the manufacturer's maximum recommended flux for that element type (e.g., typically 18-22 GFD for lead BW elements, lower for SW or fouling-resistant)? High lead-element flux leads to rapid fouling. Also check the flux of the last element – is it reasonably high (e.g., > 6-8 GFD)? Very low tail-end flux indicates poor utilization.
5. **Pressure Drop (ΔP):** Check the ΔP across individual elements (should generally be low, < 15 psi), across each pressure vessel (sum of element ΔPs, < 50-60 psi often targeted), and across each stage. High ΔP indicates potential flow issues or need for different feed spacers.
6. **Cross-Flow Velocity:** Check the minimum feed channel velocity, particularly in the last elements of the final stage. Is

it above the minimum recommended by the manufacturer (depends on spacer thickness and fouling potential)?

7. **Warnings/Flags:** Address any warnings explicitly flagged by the software. They are there for a reason!

7.6.3 Sanity Checks and Engineering Judgment

- Compare the required feed pressure to the estimated osmotic pressure – does the difference (representing NDP + losses) seem reasonable?
- Compare average flux to typical ranges for the application.
- Does the overall design seem robust, or is it pushing limits too aggressively? Consider building in some safety margin, especially if feed water quality is variable.
- The software is a powerful tool, but it relies on the user's inputs and assumptions. The engineer must apply their knowledge and experience to ensure the final design is practical and reliable.

7.7 Chapter Summary and Key Takeaways

- RO system design translates requirements into specifications using calculations and software simulations.
- Understanding fundamental calculations (osmotic pressure, NDP, flux, area, recovery, beta, salt balance, scaling indices, ΔP) provides insight and allows for sanity checks.
- Design is iterative, balancing performance, cost, and reliability.
- Manufacturer design software (e.g., WAVE, IMSDesign) is essential for accurate and efficient design of complex systems, incorporating proprietary membrane data.
- Accurate, complete, and recent feed water analysis is the non-negotiable foundation for any reliable design.
- Key design parameters include permeate flow target, membrane selection, system configuration (array, passes, etc.), target recovery, and pretreatment assumptions (pH, antiscalant).
- Critically interpreting software outputs is paramount. Pay closest attention to:

  - **Scaling potential:** Ensure all indices are within safe limits.
  - **Flux limits:** Avoid excessive flux in lead elements.
  - Permeate quality targets.
  - Hydrodynamic limits (ΔP, cross-flow).
- Software is a tool; engineering judgment and experience are required to produce a robust and reliable final design.

7.8 References

1. DuPont Water Solutions. WAVE (Water Application Value Engine) Software and Documentation. (Example of manufacturer software).
2. Hydranautics (A Nitto Group Company). IMSDesign (Integrated Membrane Solutions Design) Software and Documentation. (Example of manufacturer software).
3. Toray. TorayDS2 Software and Documentation. (Example of manufacturer software).
4. LG Chem Water Solutions. Q+ Projection Software and Documentation. (Example of manufacturer software).
5. AWWA Manual M46: Reverse Osmosis and Nanofiltration (2007). American Water Works Association. (Chapter on System Design).
6. Antiscalant Manufacturer Technical Data Sheets and Software (e.g., from Avista, Nalco Water, Kemira, Suez WTS): Provide specific limits for their products under various conditions.
7. Voutchkov, N. (2018). Desalination Engineering: Planning and Design. McGraw-Hill Education. (Practical guidance on design inputs and considerations).
8. Malek, A., Hawlader, M. N. A., & Ho, J. C. (1996). Design and costing of reverse osmosis seawater desalination plants. Desalination, 107(2), 117-131. (Example illustrating design parameters and cost factors).

Chapter 8: Water Quality Guidelines - Setting the Stage for Success

8.1 Introduction: Why Pretreatment is Non-Negotiable

In the preceding chapters, we've marveled at the elegance of the Reverse Osmosis process – its ability to separate dissolved salts from water using sophisticated membranes and carefully designed systems. However, it's crucial to remember what RO membranes are designed for: removing dissolved constituents at the molecular or ionic level. They are not designed to handle raw, untreated water straight from a river, well, or the sea.

Imagine using the finest silk cloth to filter muddy water. It might work for a moment, but it would quickly become hopelessly clogged and useless. Similarly, RO membranes, particularly the common spiral-wound TFC type with their tightly packed layers and narrow feed channels, are extremely sensitive to contaminants present in typical raw water sources. Sending unprepared water directly to an RO system is a recipe for disaster, leading to rapid performance decline, irreversible damage, frequent and costly cleaning, shortened membrane life, and overall operational failure.

Pretreatment is the series of steps applied to the raw water before it reaches the RO membranes, designed to remove or modify substances that could harm the membranes or compromise the system's performance. It's not optional; it's an integral and critical part of any successful RO plant. The extent and type of pretreatment required depend entirely on the quality of the raw water source, but the goal is always the same: to deliver water to the RO membranes that meets specific quality guidelines.

This chapter focuses on defining those guidelines. What constitutes "good quality" feed water for an RO membrane? We'll examine the key categories of contaminants that cause problems and discuss the target levels generally considered acceptable for ensuring reliable, long-term RO operation. Achieving these targets is the objective of the pretreatment technologies discussed in the next chapter.

8.2 The Clogging Culprits: Suspended Solids & Colloids

Perhaps the most intuitive challenge is dealing with undissolved particles floating in the water. These can range from visible grit and sand down to microscopic clays and organic debris.

8.2.1 Defining the Problem: TSS, Colloids, and Turbidity

- **Total Suspended Solids (TSS):** This refers to particles that are large enough to be trapped by a standard laboratory filter (typically around 1.2 μm). While RO pretreatment aims to remove virtually all TSS, the real challenge for RO membranes lies in the finer material.
- **Colloids:** These are extremely small particles, typically ranging from 0.001 μm to 1 μm in size. They often remain dispersed in water due to their small size and surface charges, making them difficult to remove by simple settling or coarse filtration. Examples include clays, silica colloids, fine organic fragments, and iron or aluminum hydroxides. Colloids are notorious foulants for RO membranes.
- **Turbidity:** This is a measure of the water's cloudiness or haziness, caused by suspended and colloidal particles scattering light. It's measured in Nephelometric Turbidity Units (NTU). While turbidity gives a general indication of particulate contamination, it doesn't perfectly correlate with the membrane fouling potential, as different types of particles scatter light differently and have different "stickiness." However, low turbidity is always desirable. Generally, the target turbidity for RO feed water (after pretreatment) is very low, often < 0.1 - 0.2 NTU.

8.2.2 The Gatekeeper Metric: Silt Density Index (SDI)

Because turbidity alone isn't a reliable predictor of fouling by fine particles, the Silt Density Index (SDI) was developed specifically for RO applications (as discussed in Chapter 3). It measures the plugging rate of a 0.45 μm filter pad under standard pressure, providing a direct, albeit empirical, assessment of the fouling potential from particulates and colloids in that critical size range.

- **The Guideline:** Virtually all RO membrane manufacturers specify a maximum SDI15 (measured over 15 minutes) for their feed water, typically SDI15 < 5.0, with a strong recommendation to maintain SDI15 < 3.0 for optimal performance and warranty validity, especially for TFC membranes.
- **Interpretation:** An SDI value consistently below 3 suggests the pretreatment system is effectively removing the vast majority of problematic fine particles and colloids. Values creeping up towards 5 are a warning sign, and values above 5 indicate a high risk of rapid particulate fouling.

8.2.3 Operational Consequences of High Solids

What happens if water with high SDI reaches the membranes?

- **Physical Blockage:** Fine particles wedge into the narrow feed spacer channels of spiral-wound modules, obstructing flow paths.
- **Surface Coating:** Particles deposit on the membrane surface, forming a cake layer.
- **Increased Differential Pressure (ΔP):** The blockage of feed channels significantly increases the pressure drop across the pressure vessel and the entire stage. This is often the first and clearest sign of particulate fouling.
- **Reduced Permeate Flux:** The cake layer adds resistance to water flow, and blocked channels reduce effective membrane area, both leading to lower water production.

- **Concentration Polarization Effects:** The fouling layer can hinder the back-diffusion of salts, worsening concentration polarization and potentially reducing salt rejection.
- **Abrasion:** Larger, harder particles (if they get past the cartridge filter) can potentially abrade pump components or even the membrane surface over time.
- **Difficult Cleaning:** While chemical cleaning can remove some particulate fouling, heavily compacted layers or deeply plugged channels can be very difficult to clean effectively, sometimes requiring aggressive (and potentially damaging) procedures or even premature membrane replacement.

Practical Insight: Religiously monitoring the SDI after the final pretreatment step (usually after the cartridge filters) is paramount. It's the RO system's early warning system for particulate threats. Ignoring rising SDI values is like ignoring the smoke detector in your house – trouble is likely brewing.

8.3 The Slime Factor: Microbes and Biofouling

Water is life, and unfortunately, that means it's also a potential home for microorganisms – bacteria, fungi, algae. While RO membranes effectively reject these microbes (they are much larger than the membrane's "pores"), their presence in the feed water leads to one of the most persistent and problematic issues in RO operation: Biofouling.

8.3.1 The Invisible Invaders: Bacteria, Fungi, Algae

Various microbes can thrive in water systems, especially if nutrients are present and conditions (temperature, pH) are favorable. Even seemingly clean water can contain enough bacteria to initiate biofouling.

8.3.2 Biofilm: The Tenacious Enemy

Microbes don't just float around individually. They tend to attach to surfaces (like membranes and spacers) and establish colonies encased in a protective, slimy matrix of Extracellular Polymeric Substances (EPS). This slimy layer is called a biofilm.

- **Formation:** A few pioneering bacteria attach, multiply, and begin secreting EPS. This sticky matrix traps other microbes, nutrients, and even inert particles, allowing the biofilm community to grow and thicken.
- **Protection:** The EPS matrix shields the embedded microbes from disinfectants and cleaning agents, making established biofilms notoriously difficult to remove completely.
- **Impact on RO:**
    - **Severe Flux Decline:** The biofilm acts as a significant barrier to water flow, often causing rapid and dramatic drops in permeate production.
    - **Increased Differential Pressure (ΔP):** The thick, gelatinous biofilm can severely clog feed channels, leading to very high pressure drops across stages. ΔP increase due to biofouling can often be faster and more severe than from particulate fouling alone.
    - **Reduced Salt Rejection:** Some bacterial enzymes can potentially degrade certain membrane materials (especially older CA types, but even polyamide can be affected over long periods). More commonly, the physical presence of the biofilm can disrupt flow patterns near the membrane surface, potentially compromising the integrity of the rejection mechanism or masking underlying issues.
    - **Difficult Cleaning:** Standard cleaning chemicals often struggle to penetrate the EPS matrix and kill all embedded microbes. Biofilm removal usually requires specific biocides and multi-step cleaning protocols, and complete removal is rarely achieved – dormant cells can regrow quickly once conditions become favorable again.
    - **Downstream Contamination:** Biofilm fragments can slough off and contaminate the permeate stream, a major concern in applications requiring high biological purity (e.g., pharmaceutical water).

8.3.3 Indicators and Control Philosophy

Detecting the onset of biofouling before it becomes severe can be tricky.

Indicators:

- Rapid increase in normalized ΔP, often more pronounced than the decline in normalized flux, is a classic sign.
- Increased frequency of cartridge filter changes (biofilm sloughing off upstream).
- Presence of slime observed during membrane cleaning or autopsy is definitive proof.
- Plate counts (HPC) can indicate microbial load in feed or permeate, but don't directly measure biofilm on the membrane surface. Low counts are desirable but don't guarantee absence of biofilm.
- ATP (Adenosine Triphosphate) monitoring: Measures total biological activity, potentially offering a faster indication than plate counts.
- Microscopic analysis of feed water or deposits.

Control Philosophy:

Since complete eradication is difficult, the focus is on prevention and control:

- **Minimize Nutrient Load:** Pretreatment steps that remove organics (food source) can help limit growth potential.
- **Effective Pre-Disinfection (if applicable):** Using biocides like chlorine before dechlorination (for TFC membranes) can reduce the incoming microbial load. Careful control and complete removal of the disinfectant are essential. Alternatively, UV disinfection can be used.
- **Proper Flushing:** Performing regular low-pressure permeate flushes, especially before shutdown, removes nutrient-rich feed water from the membrane elements.
- **Avoid Stagnation:** Keep the system operating or properly flushed/preserved. Long periods of stagnation are prime opportunities for biofilm growth.
- **Periodic Sanitization/Biocidal Cleaning:** Incorporating specific biocides into the regular cleaning regimen helps keep microbial populations in check.

- **Membrane Preservation:** Using appropriate preservative solutions (e.g., sodium bisulfite solution) during long-term shutdowns.

Guidelines: There isn't a single universally accepted numerical guideline like SDI for microbial load entering RO. The goal is multifaceted: minimize incoming viable cell count, limit nutrient availability, prevent stagnant conditions, and implement proactive cleaning/disinfection strategies.

Real-life Example: A brackish water RO plant operating smoothly in winter experiences a rapid increase in differential pressure and drop in production as the weather warms up in spring. Investigation reveals significant slime on the membrane elements – classic seasonal biofouling triggered by warmer temperatures accelerating microbial growth. This necessitates an immediate, aggressive cleaning cycle using specialized biocides.

8.4 The Organic Burden: Dissolved Organics and Their Impact

Beyond living microbes, non-living organic compounds dissolved or suspended in water can also foul RO membranes.

8.4.1 Natural and Synthetic Organics

- **Natural Organic Matter (NOM):** A complex mixture of large molecules derived from the decomposition of plant and animal matter. Key components include humic acids and fulvic acids, common in surface waters (rivers, lakes) and some groundwaters. These compounds often impart color to water.
- **Synthetic Organic Chemicals:** Can originate from industrial discharges, agricultural runoff (pesticides, herbicides), or wastewater effluent (pharmaceutical residues, personal care products).
- **Oils and Greases:** Can be present due to industrial contamination or natural sources.

8.4.2 Fouling by Adsorption and Bio-Nourishment

Organics cause problems in several ways:

- **Adsorption:** NOM molecules, particularly humic substances, have an affinity for polyamide membrane surfaces and can adsorb onto them. This forms an organic fouling layer that restricts water passage, leading to flux decline.
- **Pore Blocking:** Larger organic molecules or colloidal organics might partially block membrane 'pores' or entrance points.
- **Coating:** Oils and greases can form a water-repellent film on the membrane, drastically reducing flux.
- **Biofouling Enhancement:** Many dissolved organics serve as nutrients (food) for bacteria, promoting biofilm growth. Organic fouling layers can also provide a favorable surface for microbial attachment.

8.4.3 Measuring the Load: TOC and Other Indicators

Assessing the organic fouling potential involves several measurements:

- **Total Organic Carbon (TOC):** The most common measure of the total amount of organic matter, expressed as mg/L of carbon. While a useful indicator, TOC doesn't reveal the type of organics present (some types foul more readily than others) nor their bioavailability. General guidelines often suggest keeping TOC in RO feed below 2-3 mg/L, and sometimes below 1 mg/L for surface water sources known to be high in fouling NOM species. However, these are just guidelines; the acceptable level depends greatly on the specific organics and membrane type.
- **UV Absorbance at 254 nm (UVA_{254}):** Aromatic organic compounds (common in NOM) strongly absorb UV light at this wavelength. It's a quick indicator of certain types of NOM but doesn't capture all organics.
- **Specific Organic Carbon fractions (e.g., DOC, BDOC):** Dissolved Organic Carbon (DOC) measures the soluble fraction. Biodegradable DOC (BDOC) specifically measures

the fraction that can serve as food for microbes, relevant for biofouling potential.

- **Oil & Grease Analysis:** Specific laboratory tests are needed if oil contamination is suspected. Even trace amounts can be problematic.

Practical Insight: Surface waters are often high in NOM and require significant pretreatment (e.g., coagulation, flocculation, sedimentation, media filtration, sometimes UF or activated carbon) to reduce TOC and color before RO. Groundwater is typically lower in NOM but can sometimes contain specific synthetic organics depending on local geology and contamination sources.

8.5 The Telltale Sign: Color as an Indicator

While color itself might not directly harm the RO membrane's separation ability, its presence in the feed water is often a strong indicator of potential problems, primarily related to organics.

8.5.1 What Causes Color?

True color (color remaining after filtering out suspended solids) is usually caused by dissolved organic substances, primarily tannins, humic acids, and fulvic acids originating from decaying vegetation. Think of the tea-like color of water from swampy areas or forested catchments. Less commonly, color can be caused by dissolved metals like iron or manganese, or certain industrial contaminants.

8.5.2 Why Color Matters to RO Pretreatment

- **Indicator of NOM:** High color almost always signifies high levels of Natural Organic Matter (NOM). As discussed in Section 8.4, NOM is a known membrane foulant. Therefore, if the feed water has significant color, it strongly suggests that organic fouling will be a major concern unless effective pretreatment for NOM removal is implemented.
- **Pretreatment Target:** While RO membranes generally reject the large molecules causing color very well (permeate is usually colorless), the goal of pretreatment is to remove these

color-causing organics before they reach the RO to prevent fouling. There isn't a strict color guideline for the RO feed itself if the organics causing it are successfully managed (e.g., TOC is low), but achieving low color (e.g., < 5-10 Color Units) in the pretreated water is a good sign that the upstream organic removal process is working effectively.

Real-life Example: A facility drawing water from a river known for high seasonal color due to runoff from forests must employ robust pretreatment, often involving coagulation to bind the color-causing organics into larger flocs, followed by filtration, to protect their RO membranes from rapid organic fouling during high-color periods.

8.6 Chapter Summary: The Target Water Quality Profile

The goal of pretreatment is to deliver water to the RO membranes that meets stringent quality guidelines to ensure efficient, reliable, and long-lasting operation. While specific values can vary slightly based on membrane type, system design, and operating conditions, the generally accepted targets for RO feed water (after pretreatment) are:

- Suspended Solids/Colloids: SDI15 < 3.0 (ideally), definitely < 5.0. Turbidity < 0.1 - 0.2 NTU.
- Microbes: Minimized viable count and growth potential. Requires a multi-faceted control strategy (disinfection/dechlorination, nutrient removal, flushing, cleaning) rather than a single numerical target, although low HPC (<100 CFU/mL) is desirable. Absence of biocidal residuals (e.g., low ORP for chlorine-sensitive membranes) is critical.
- Organics: TOC < 2-3 mg/L (general guideline, lower may be needed for high-fouling NOM sources). Oils & Grease essentially absent (< 0.1 mg/L or non-detect).
- Color: Low (e.g., < 5-10 CU) as an indicator of successful organic removal.
- Potential Scalants: Concentrations must be low enough such that, after considering the concentration factor (Beta) in the

RO system, scaling indices (LSI, $CaSO_4$%, SiO_2%) remain within safe limits, often requiring the use of antiscalants or pH adjustment (covered implicitly by design calculations in Chapter 7, but achieved via pretreatment techniques in Chapter 9).

- Free Chlorine/Oxidizers: Essentially zero ($< 0.05 - 0.1$ mg/L) for TFC polyamide membranes.

Meeting these targets requires careful selection and operation of appropriate pretreatment technologies, which we will explore in detail in the next chapter. Failing to meet these guidelines inevitably translates into higher operating costs, more frequent downtime, and a shorter lifespan for the heart of the system – the RO membranes.

8.7 References

1. AWWA Manual M46: Reverse Osmosis and Nanofiltration (2007). American Water Works Association. (Chapter on Feedwater Quality Considerations).
2. DuPont Water Solutions. FilmTec™ Membranes Technical Manual. (Provides specific feed water limits and guidelines for their membranes, including SDI, chlorine tolerance, etc.).
3. Hydranautics (A Nitto Group Company). Technical Service Bulletins. (Often contain detailed feed water quality recommendations).
4. ASTM Standard D4189-07(2014). Standard Test Method for Silt Density Index (SDI) of Water. ASTM International.
5. Ridgway, H. F., & Flemming, H. C. (1996). Biofouling of membranes. In Mallevialle, J., Odendaal, P. E., & Wiesner, M. R. (Eds.), Water Treatment Membrane Processes (pp. 6.1-6.61). McGraw-Hill. (Comprehensive overview of biofouling mechanisms).
6. Flemming, H. C., Schaule, G., & McDonogh, R. (1994). How do performance parameters respond to initial biofouling on separation membranes?. Vom Wasser, 83, 183-192. (Classic work on biofouling impact).
7. Schäfer, A. I., Fane, A. G., & Waite, T. D. (Eds.). (2005). Nanofiltration: Principles and Applications. Elsevier. (Chapters discuss fouling by organics and NOM).

8. Hong, S., & Elimelech, M. (1997). Chemical and physical aspects of natural organic matter (NOM) fouling of nanofiltration membranes. Journal of Membrane Science, 132(2), 159-181. (Detailed study on NOM fouling mechanisms).

Chapter 9: Pretreatment Techniques and Technologies - The RO Preparation Toolkit

9.1 Introduction: Building the Pretreatment Train

Meeting the stringent feed water quality guidelines outlined in Chapter 8 rarely involves a single treatment step. Instead, a sequence of technologies, known as the **pretreatment train**, is carefully selected and arranged to progressively remove or modify problematic contaminants. The complexity of this train depends heavily on the raw water source: a relatively clean groundwater source might only require cartridge filtration and antiscalant addition, whereas a challenging surface water source might necessitate screening, coagulation, flocculation, sedimentation, media filtration, cartridge filtration, dechlorination, and antiscalant dosing.

This chapter explores the individual unit processes commonly employed in RO pretreatment trains. We'll examine how each technology works, what contaminants it targets, and its typical place within the overall sequence. Understanding this toolkit is essential for designing effective pretreatment systems and troubleshooting performance issues.

9.2 Physical Removal - Part I: Basic Filtration

The first objective is often the removal of larger suspended particles.

9.2.1 Coarse Screening:

Purpose: To remove large debris like leaves, twigs, rags, fish, and grit that could damage downstream equipment, particularly pumps.

Mechanism: Simple physical barriers like bar screens (parallel bars) or travelling band screens (rotating mesh screens).

Application: Primarily used for surface water intakes (rivers, lakes, seawater). Usually considered part of the intake structure rather than the main pretreatment skid, but essential protection.

9.2.2 Media Filtration: The Conventional Workhorse

For decades, granular media filters have been the standard method for removing finer suspended solids than screens can capture.

9.2.2.1 Mechanisms:

Water flows downward through a bed of granular media. Removal occurs through several mechanisms simultaneously:

- **Straining:** Particles larger than the pore spaces between media grains are physically trapped, mostly at the surface.
- **Interception & Impaction:** Smaller particles following flow lines collide with and adhere to media grains.
- **Adsorption:** Physical and chemical attraction between particles and the media surface.
- **Sedimentation:** Some settling occurs within the pores.
- **Biological Action:** Mature filters can develop a biological layer that helps trap particles (though excessive biological growth is undesirable).

9.2.2.2 Filter Media: Different materials can be used, often in layers:

- **Single Media (Sand Filters):** Use sand of a relatively uniform size. Effective for water with low to moderate solids loading.
- **Dual Media:** A coarser top layer of low-density anthracite coal sits above a finer layer of denser sand. Allows deeper penetration of solids into the bed before surface plugging occurs (depth filtration), leading to longer filter runs between backwashes.
- **Multimedia:** Typically includes a top layer of anthracite, a middle layer of sand, and a bottom layer of very fine, very dense garnet sand. Sometimes includes a supporting gravel layer. This graded density approach optimizes depth filtration,

allowing for higher filtration rates and longer runs. This is very common in RO pretreatment.

9.2.2.3 Operation: Service Cycle and Backwashing:

- **Service Cycle:** Water flows downward through the media until either the pressure drop (ΔP) across the bed reaches a set limit (e.g., 8-10 psid) due to trapped solids, or a maximum time has elapsed, or sometimes when outlet turbidity increases (breakthrough).
- **Backwashing:** Flow is reversed, sending clean water (often filtered water from other units or storage) upward through the media bed at a high enough velocity to fluidize the bed, expand it, and flush the trapped solids out to waste. Often preceded by an air scour step (injecting air) to help break up compacted solids. Proper backwashing is critical for maintaining filter performance. Ineffective backwashing is a common cause of filter failure (mudball formation, channeling).

9.2.2.4 Performance and Limitations:

Well-operated media filters can typically reduce turbidity significantly and produce water with an SDI15 often in the 4-6 range. However, performance can be variable, especially with changes in raw water quality. They are less effective at removing very fine colloids without chemical assistance (coagulation). Breakthrough of fine particles ("fines") can occur, especially immediately after backwashing, potentially stressing downstream cartridge filters.

9.3 Physical Removal - Part II: Enhanced Floc Removal

Simple media filtration often struggles with the very fine colloidal particles (< 1 μm) that significantly contribute to RO membrane fouling potential (high SDI). To tackle these, chemical destabilization followed by physical separation is often employed, particularly for surface waters.

9.3.1 The Need:

Colloids typically carry a negative surface charge, causing them to repel each other and remain stably suspended. Physical filtration alone may not capture them effectively.

9.3.2 Coagulation and Flocculation:

Making Small Particles Big. This two-step process uses chemicals to destabilize colloids and promote their aggregation into larger, settleable or filterable clumps called "flocs."

9.3.2.1 Coagulation:

Charge Neutralization: A coagulant chemical, typically a positively charged metal salt or polymer, is added and rapidly mixed with the water (flash mix). Common coagulants include:

- **Aluminum Sulfate (Alum):** Widely used, effective, relatively inexpensive. Works best in a specific pH range (e.g., 6-7.5). Produces aluminum hydroxide floc. Residual aluminum can be a foulant if carryover occurs.
- **Ferric Chloride / Ferric Sulfate:** Effective over a broader pH range than alum. Produces dense ferric hydroxide floc. Can impart slight color if residual iron carries over.
- **Polymers (e.g., PolyDADMAC, Polyacrylamide variants):** Long-chain organic molecules that can neutralize charge and bridge particles. Often used as primary coagulants or as "coagulant aids" along with metal salts to strengthen flocs and reduce metal salt dose. Careful selection and dosing are crucial.

The goal is to neutralize the negative charges on the colloidal particles, overcoming repulsive forces and allowing them to collide and stick together upon contact (van der Waals forces).

9.3.2.2 Flocculation:

Gentle Mixing for Growth: After rapid mixing of the coagulant, the water flows into a basin with slow, gentle mixing (using paddle mixers or baffles). This provides time and opportunity for the destabilized micro-flocs to collide and aggregate into larger, visible flocs. Mixing energy must be carefully controlled – too little and flocs won't form effectively, too much and flocs can be sheared apart.

9.3.2.3 Jar Testing:

Optimizing Chemical Doses: Determining the right type and dose of coagulant (and flocculant aid, if used) and the optimal pH is critical and highly dependent on the specific water chemistry (turbidity, alkalinity, temperature, organic content). Jar testing is the standard laboratory procedure used to simulate coagulation/flocculation/settling in beakers to find the optimal chemical conditions before applying them at full scale.

9.3.3 Sedimentation (Clarification): Letting Gravity Work

Purpose: After flocs have formed, this step provides a period of quiescence in a large basin (clarifier) to allow the heavier-than-water flocs to settle out by gravity.

Mechanism: Water flows slowly through the clarifier, allowing flocs to settle to the bottom as sludge, which is periodically removed. Clearer water overflows from the top.

Application: Common in conventional surface water treatment plants preceding media filtration. Reduces the solids load going to the filters, allowing them to operate more efficiently and for longer periods.

9.3.4 Dissolved Air Flotation (DAF): Floating the Flocs

Purpose: An alternative to sedimentation, particularly effective for low-density flocs (e.g., those formed when treating algae-laden water or low-turbidity, colored water) that don't settle well.

Mechanism: A portion of the clarified effluent is recycled, saturated with air under pressure, and then mixed with the flocculated water entering the DAF tank at atmospheric pressure. The sudden pressure release creates millions of microscopic air bubbles that attach to the flocs, making them buoyant. The floc-bubble aggregates rise to the surface, forming a sludge blanket ("float") that is skimmed off. Clarified water is withdrawn from the bottom.

Application: Increasingly popular for challenging surface waters and seawater pretreatment due to its effectiveness with difficult flocs and often smaller footprint compared to sedimentation basins.

Overall Impact: Coagulation/flocculation followed by sedimentation or DAF significantly reduces turbidity and the concentration of colloidal particles, greatly improving the performance and reducing the load on downstream filters (media or membrane).

9.4 Adsorption and Oxidation Control: Activated Carbon Filtration

Activated carbon is a highly porous material with an enormous internal surface area, making it excellent at adsorbing certain types of dissolved molecules from water.

9.4.1 Mechanism:

Adsorption is a surface phenomenon where molecules (adsorbates) adhere to the surface of a solid (adsorbent – the activated carbon) due to physical forces (van der Waals) or sometimes chemical interactions.

9.4.2 Targets:

- **Dissolved Organic Matter (NOM):** Particularly effective at removing certain types of NOM that cause taste, odor, and color, and can contribute to organic fouling of RO membranes.
- **Synthetic Organic Chemicals:** Can remove pesticides, industrial solvents, etc.
- **Chlorine/Chloramine Removal (Dechlorination):** Activated carbon catalytically destroys free chlorine (hypochlorous acid) and chloramines, converting them primarily to chloride ions. This is a critical application when using chlorine-intolerant TFC membranes. It acts both as an adsorbent and a chemical reactant in this case.

9.4.3 Considerations:

- **Type of Carbon:** Granular Activated Carbon (GAC) is used in filter beds similar to media filters. Powdered Activated Carbon (PAC) can be added earlier in the process (e.g., during coagulation) but is less common for RO pretreatment. GAC can be made from coal, coconut shells, or wood; different source materials yield carbons with different pore structures and adsorption characteristics.
- **Contact Time (EBCT):** Effective adsorption requires sufficient contact time between the water and the carbon, typically expressed as Empty Bed Contact Time (EBCT - the theoretical time water spends in the filter bed). Longer EBCT (e.g., 10-20 minutes) is needed for significant organic removal, while shorter EBCT (e.g., 3-5 minutes) may suffice for dechlorination.
- **Exhaustion:** The adsorption capacity of the carbon is finite. Eventually, the sites become saturated, and the carbon needs to be replaced or reactivated (a high-temperature process usually done off-site). Exhaustion for dechlorination happens much more slowly than for organic removal. Monitoring downstream chlorine levels (or ORP) is crucial if GAC is used for dechlorination.
- **Biological Growth:** GAC beds can become breeding grounds for bacteria since they remove disinfectant and adsorb nutrients. Regular backwashing is needed, and the potential for bacteria sloughing off into the RO feed must be considered.

9.5 Chemical Conditioning: Addressing Scaling and Microbes

These steps don't physically remove contaminants but chemically modify the water to prevent problems within the RO system itself.

9.5.1 Antiscalant Addition: Preventing Mineral Precipitation

Purpose: To inhibit the formation of mineral scale ($CaCO_3$, $CaSO_4$, $BaSO_4$, $SrSO_4$, sometimes silica) on the membrane surface, even when the concentrate stream becomes supersaturated. This allows RO systems to operate at higher recovery rates than would otherwise be possible.

Mechanism: Antiscalants (also called scale inhibitors) are specialty chemicals, typically phosphonates or proprietary polymer blends. They work at sub-stoichiometric concentrations (a few ppm) through several mechanisms:

- Threshold Inhibition: Interfere with the initial formation of scale crystals (nucleation).
- Crystal Modification: Distort the structure of growing crystals, making them less likely to adhere to surfaces or form hard scale.
- Dispersion: Adsorb onto crystals or colloidal particles, increasing their negative charge and causing them to repel each other, keeping them dispersed.

Selection & Dosing: The correct antiscalant product and dose rate must be selected based on a detailed water analysis, the RO system operating conditions (recovery, temperature), and the specific scaling potential predicted by software (LSI, $CaSO_4$%, etc.). Vendor recommendations are crucial. Over- or under-dosing can be ineffective or even detrimental. Dosed continuously using accurate metering pumps just before the cartridge filters.

9.5.2 Acid Addition (pH Adjustment): Controlling Carbonate Scale

Purpose: Primarily to prevent Calcium Carbonate ($CaCO_3$) scaling by lowering the feed water pH.

Mechanism: $CaCO_3$ solubility is highly dependent on pH. Adding an acid (commonly sulfuric acid H_2SO_4, sometimes hydrochloric acid

HCl) converts bicarbonate (HCO_3^-) and carbonate (CO_3^{2-}) ions into dissolved carbonic acid (H_2CO_3) / CO_2, significantly increasing $CaCO_3$ solubility and lowering the LSI. Typically, pH is lowered to around 6.0-6.5.

Pros & Cons: Effective for $CaCO_3$ control. Can be less expensive than antiscalants in some cases. However:

- Requires careful handling of corrosive acids.
- Precise pH control is needed (over-acidification increases corrosion risks).
- Does not prevent sulfate or silica scaling.
- Increases the amount of dissolved CO_2 in the feed, which readily passes through the RO membrane, lowering permeate pH and increasing permeate conductivity (CO_2 forms carbonic acid in the permeate). This may require downstream degassing or pH adjustment of the permeate.

Application: Less common as the primary scaling control method in large modern systems compared to antiscalants, but still used in some specific applications or smaller systems, sometimes in combination with a reduced antiscalant dose.

9.5.3 Biocide Addition & Dechlorination: Managing Microbial Loads

Controlling microbial growth is crucial throughout the pretreatment and RO system.

9.5.3.1 Pre-Disinfection Strategies:

Applying a disinfectant early in the pretreatment train can reduce the overall microbial load reaching downstream components.

- **Chlorination:** Adding chlorine (gas or hypochlorite) is common in municipal water treatment. It effectively kills bacteria but must be removed before TFC membranes. Requires sufficient contact time. Can form disinfection byproducts (DBPs).

- **Chloramination:** Using chloramines (formed by adding ammonia after chlorine) provides a more stable, longer-lasting residual disinfectant but is less potent than free chlorine. TFC membranes have limited tolerance to chloramines (check manufacturer specs). Also needs removal/reduction before RO.
- **Ultraviolet (UV) Disinfection:** Uses UV light (typically at 254 nm) to inactivate microorganisms by damaging their DNA. Does not add chemicals and doesn't produce DBPs. Often used after media filtration and before RO as a non-chemical disinfection step, or sometimes earlier. Does not provide a residual disinfectant effect.
- **Non-oxidizing Biocides:** Specialty chemicals sometimes dosed intermittently to control specific microbial issues, but require careful selection for compatibility and environmental considerations.

9.5.3.2 The Critical Step: Dechlorination:

If chlorine or chloramine is used upstream, its removal before TFC membranes is absolutely mandatory.

- **Sodium Bisulfite (SBS) Injection:** Most common method. SBS ($NaHSO_3$) is a reducing agent rapidly injected just before the cartridge filters. It chemically reacts with free chlorine and chloramines, converting them to harmless chloride and sulfate ions. Requires accurate dosing based on the residual chlorine level, plus a slight excess. ORP monitoring downstream confirms removal (target ORP typically < 150-200 mV).
- **Granular Activated Carbon (GAC) Filtration:** As mentioned earlier, GAC beds effectively remove chlorine/chloramines through catalytic reaction. Provides an alternative or backup to SBS dosing. Requires monitoring for carbon exhaustion.

9.6 Advanced Barrier: Membrane Filtration Pretreatment

For applications requiring exceptionally high-quality RO feed water, particularly when dealing with challenging surface waters or wastewater reuse, membrane filtration technologies like Microfiltration (MF) or Ultrafiltration (UF) are increasingly used as pretreatment.

9.6.1 Microfiltration (MF):

Precise Particle Removal

- **Pore Size:** Typically 0.1 to 0.4 µm range.
- **Mechanism:** Primarily size exclusion. Operates at low transmembrane pressure (typically 5-30 psi).
- **Targets:** Effectively removes suspended solids, algae, protozoa (like Giardia, Cryptosporidium), and bacteria. Does not significantly remove viruses or dissolved molecules.
- **Result:** Produces water with very low turbidity and good SDI, often < 3.

9.6.2 Ultrafiltration (UF):

Superior Colloid and Pathogen Barrier

- **Pore Size:** Typically 0.01 to 0.1 µm range (finer than MF).
- **Mechanism:** Size exclusion. Operates at slightly higher pressure than MF (typically 10-50 psi).
- **Targets:** Removes everything MF removes, plus viruses, colloids, and larger organic molecules (like humic substances, proteins). Essentially provides a near-complete physical barrier for particulates and microbes.
- **Result:** Produces exceptionally high-quality water with consistently low turbidity (< 0.1 NTU) and consistently low SDI (< 1-2), ideal for feeding RO membranes. Significantly reduces fouling potential from particulates, colloids, and microbes.

9.6.3 Advantages for RO Pretreatment:

- **Consistent Effluent Quality:** Less susceptible to upsets from changes in raw water quality compared to conventional media filters + clarification.
- **Absolute Barrier:** Provides reliable removal of pathogens and fine particles.
- **Reduced Chemical Use:** May reduce or eliminate the need for coagulants compared to conventional treatment achieving similar SDI.
- **Compact Footprint:** Often requires less space than large clarifiers and media filters.

9.6.4 Configurations and Considerations:

Modules: Commonly use hollow fiber membranes (either inside-out or outside-in flow) or sometimes spiral-wound or tubular configurations.

Operation: Typically operate in cross-flow or semi-dead-end mode with periodic backwashing (often chemically enhanced with chlorine, acid, or caustic – UF membranes are generally more robust than RO membranes) to remove accumulated solids.

Cost: Higher capital cost than conventional media filtration, but can lead to lower overall operating costs for the RO system due to reduced fouling, cleaning frequency, and potentially longer RO membrane life. Requires membrane integrity testing to ensure effectiveness.

9.7 Selecting the Right Combination: Factors Influencing Choice

The optimal pretreatment train is site-specific and depends on:

1. **Raw Water Source Quality:** The biggest factor (Surface vs. Ground vs. Seawater vs. Wastewater; levels of TSS, colloids, organics, microbes, salinity, specific ions).

2. **RO Membrane Type:** Primarily TFC (chlorine intolerant) vs. older CA (some tolerance).
3. **Required RO Performance:** Target recovery rate (higher recovery demands better pretreatment, especially scaling control).
4. **Reliability Requirements:** Critical applications may justify more robust (and expensive) pretreatment like UF.
5. **Capital Costs:** Conventional methods are often cheaper initially than membrane pretreatment.
6. **Operating Costs:** Chemical consumption, power usage, labor, sludge/waste disposal.
7. Footprint/Space Availability.
8. Operator Skill Level.

Often, a combination of techniques provides the best balance of performance, reliability, and cost.

9.8 Chapter Summary and Key Takeaways

- Pretreatment is essential to protect RO membranes and ensure reliable operation. The goal is to meet the feed water quality guidelines established in Chapter 8.
- A pretreatment train uses multiple technologies in sequence.
- Basic Filtration (Screening, Media Filters) removes larger suspended solids. Multimedia filters offer depth filtration. Backwashing is critical.
- Enhanced Floc Removal (Coagulation, Flocculation, Sedimentation/DAF) uses chemicals to aggregate and remove fine colloids, common for surface water. Jar testing optimizes chemicals.
- Activated Carbon (GAC) adsorbs organics and catalytically removes chlorine/chloramines, protecting TFC membranes.
- Chemical Conditioning includes Antiscalants (inhibits scaling), Acid (lowers pH for $CaCO_3$ control), Biocides (controls microbes), and Dechlorination (removes chlorine via SBS or GAC).

- Membrane Pretreatment (MF/UF) provides a superior physical barrier, producing consistently low SDI/turbidity water ideal for RO feed, often reducing RO fouling potential significantly.
- The optimal pretreatment strategy depends heavily on raw water quality, RO system requirements, and economic factors, requiring careful engineering analysis.

9.9 References

1. AWWA Manual M46: Reverse Osmosis and Nanofiltration (2007). American Water Works Association. (Chapter on Pretreatment).
2. AWWA. Water Quality and Treatment: A Handbook on Drinking Water (6th ed.). (Comprehensive text covering conventional treatment processes like coagulation, filtration, activated carbon).
3. Crittenden, J. C., Trussell, R. R., Hand, D. W., Howe, K. J., & Tchobanoglous, G. (2012). MWH's Water Treatment: Principles and Design (3rd ed.). John Wiley & Sons. (Detailed engineering textbook on water treatment technologies).
4. Voutchkov, N. (2018). Desalination Engineering: Planning and Design. McGraw-Hill Education. (Covers pretreatment specific to desalination, including DAF and UF).
5. Dow/DuPont, Hydranautics, Toray, etc. (Manufacturer Literature): Often provide guidance on recommended pretreatment based on feed water types for their membranes.
6. Leiknes, T., & Ødegaard, H. (2007). The development of combined chemical and biological processes for the removal of organic matter in drinking water treatment. Water Science and Technology: Water Supply, 7(1), 139-147. (Discusses organic removal).
7. Pearce, G. K. (2007). UF/MF pre-treatment to RO in seawater and wastewater reuse applications: a comparison of energy costs. Desalination, 202(1-3), 287-295. (Highlights UF as RO pretreatment).
8. Antiscalant Manufacturer Technical Literature (Avista, Nalco, Kemira, Suez, etc.): Detail the application and selection of their specific products.

Chapter 10: RO System Operation - Keeping the Wheels Turning

10.1 Introduction: From Standstill to Production

A well-designed and properly installed RO system represents a significant investment. The ultimate return on that investment depends not just on the design, but critically on how the system is operated and maintained day-in and day-out. Careless operation can quickly lead to damaged membranes, reduced efficiency, poor water quality, and costly repairs, negating the benefits of even the most sophisticated design.

This chapter focuses on the practical "how-to" of running an RO system. We'll cover the essential steps involved in safely starting the system for the first time (commissioning), the routine checks and monitoring required during normal operation, the importance of tracking performance trends using normalized data, and the correct procedures for shutting the system down, whether for short periods or extended storage. These procedures are designed to protect the heart of the system – the membranes – and ensure the plant delivers the required quantity and quality of water reliably and efficiently. While specific steps may vary slightly depending on the system size, manufacturer recommendations, and level of automation, the underlying principles remain consistent.

10.2 Before You Begin: Pre-Start-up Checks

Whether starting a brand-new system or restarting after maintenance or a long shutdown, a thorough pre-start checklist is essential to prevent damage and ensure a smooth start-up. Rushing this stage can lead to costly mistakes.

10.2.1 Mechanical and Electrical Verification:

- Confirm all construction/maintenance work is complete and tools/debris removed.
- Check pump and motor alignment and lubrication.
- Verify pump rotation direction (bump start motor briefly).
- Ensure all guards and safety features are in place.
- Verify electrical connections are secure and panels are closed.
- Confirm power supply is available and correct voltage/phase.
- Check control system power and communication links.

10.2.2 Pretreatment System Readiness:

- Verify the pretreatment system (media filters, UF, chemical dosing, etc.) is operational and producing water that meets the RO feed specifications (especially SDI, turbidity, chlorine/oxidant levels). Never feed out-of-spec water to the RO membranes.
- Ensure chemical dosing tanks (antiscalant, SBS, pH adjust) are filled with the correct chemicals and dosing pumps are primed and calibrated.
- Confirm cartridge filters have new or clean cartridges installed.

10.2.3 Valve Positions and Calibration Checks:

- Walk the system and verify all manual and automatic valves are in their correct pre-start positions (e.g., feed inlet closed, concentrate fully open, permeate closed or diverted to drain, flush valves closed). Consult the system's Operating Manual or P&ID.
- Check that critical instruments (pressure transmitters, flow meters, conductivity sensors, ORP) are calibrated and providing plausible readings at zero flow/pressure.

10.2.4 Flushing and Venting:

- If the system is dry (new installation or after maintenance involving opening pipes/vessels), it needs to be filled slowly with water (preferably permeate or pretreated feed water at low pressure) to displace air.

- Open high-point vents on filter housings and potentially pressure vessel end caps (if available) to allow trapped air to escape. Air pockets compressing rapidly under high pressure can cause water hammer and damage components. Close vents once water flows steadily.
- Flush the feed piping up to the RO inlet valve to remove any debris from installation/maintenance.

Practical Insight: Treat this checklist seriously. Skipping steps, especially verifying pretreatment readiness and valve positions, is a common cause of start-up problems and membrane damage. For instance, starting the high-pressure pump with the concentrate valve closed can instantly over-pressurize and rupture membranes or vessels. Feeding chlorinated water to TFC membranes will cause irreversible damage within minutes.

10.3 Commissioning and Initial Start-up Procedures

Starting a new system (or one with new membranes) requires particular care, as the membranes need proper wetting and conditioning, and baseline performance data must be established.

10.3.1 Initial Wetting and Low-Pressure Flush:

- New membrane elements are often shipped with a preservative solution (like sodium bisulfite or glycerine). This needs to be flushed out.
- Slowly introduce pretreated feed water (or permeate, if available) into the system at low pressure (< 50-60 psi) and low flow, with concentrate and permeate valves open to drain.
- Flush for a specified period (e.g., 30-60 minutes or until preservative is cleared – check conductivity/odor of drain water) to thoroughly wet the membranes and flush out air and preservative. Refer to the membrane manufacturer's specific recommendations.

10.3.2 Loading Preservatives (If applicable):

This step is usually for restarting after a preserved shutdown, not initial commissioning. If membranes were stored in preservative, this solution needs to be flushed out similarly to the initial wetting step before applying high pressure.

10.3.3 Gradual Pressurization and Flow Ramp-up:

- **Critical Step:** Avoid sudden pressure surges or high flow impacts on the membranes, which can cause "telescoping" (slippage of membrane leaves) or damage.
- Slowly close the concentrate valve partially (never fully!) and slowly start the high-pressure feed pump, preferably using a VFD to ramp up speed gradually over several minutes. If no VFD, use a downstream control valve to slowly increase pressure.
- Increase feed pressure and flow incrementally towards the design operating points, carefully monitoring pressures, flows, and permeate conductivity. Allow conditions to stabilize at each step.

10.3.4 Establishing Target Parameters (Pressure, Flow, Recovery):

- Once stable operation is achieved near design conditions, fine-tune the feed pressure (via VFD or control valve) and concentrate valve position to achieve the target permeate flow rate and design recovery percentage.
- Remember that initial performance may differ slightly from design projections due to actual feed temperature and initial membrane compaction.

10.3.5 Initial Performance Data Collection (Baseline):

- Once the system is stable at design conditions, meticulously record all operating data (flows, pressures, temperature, conductivity, pH, ORP, chemical dose rates).

- Calculate and record the initial normalized flux, normalized salt passage (or rejection), and differential pressures.
- This initial data set is the crucial baseline against which all future performance will be compared to assess fouling and membrane health. Keep this record safe!

10.3.6 Permeate Diversion (Initial Quality):

- The permeate produced during the initial start-up and stabilization period may not meet the final quality requirements (due to residual preservative, initial membrane wetting effects, or temporary passage of CO_2).
- Most systems have a "Permeate Divert" or "Product Dump" valve that automatically sends the initial permeate flow to drain for a set time period (e.g., 30-60 minutes) or until conductivity drops below a target setpoint, before directing it to the service/storage tank.

Practical Insight: The first few hours and days of operation are critical. Monitor the system closely, log data frequently, and be prepared to make minor adjustments. Don't push the system to maximum design parameters immediately; allow membranes to condition. Document everything.

10.4 Routine Operation: Daily Tasks and Monitoring

Once the system is running stably, routine monitoring by the operator is essential to ensure continued performance and detect potential issues early.

10.4.1 The Operator's Log Sheet/Digital Interface:

Whether using paper log sheets or a digital HMI/SCADA system, data should be recorded systematically at regular intervals (e.g., once per shift, every few hours, or continuously logged). Consistency in recording units and conditions is key.

10.4.2 Key Parameter Checks (Flows, Pressures, Quality):

- **Flows:** Permeate Flow (main output), Concentrate Flow, Feed Flow (often calculated or measured). Check against targets. Gradual decrease in permeate flow (when normalized) indicates fouling.
- **Pressures:** Feed Pressure, Concentrate Pressure, Permeate Pressure, Differential Pressure (ΔP) across cartridge filter, ΔP across each RO stage (or vessel, if instrumented). Increasing Feed Pressure needed to maintain flow, or increasing Stage ΔP, signals fouling.
- **Quality:** Permeate Conductivity/TDS (CRITICAL indicator of rejection performance). Feed Conductivity/TDS. A gradual or sudden increase in permeate conductivity requires immediate investigation.

10.4.3 Pretreatment System Monitoring:

- **SDI:** Perform SDI tests on the RO feed water regularly (e.g., daily or per shift). An increasing trend signals pretreatment problems.
- **Cartridge Filter ΔP:** Monitor the pressure drop across the cartridge filter housing. Replace cartridges promptly when ΔP limit is reached. Log replacement frequency.
- **Chemical Levels & Dose Rates:** Check levels in antiscalant, SBS, and other chemical tanks. Verify metering pumps are operating correctly and dose rates are accurate (using calibration column draw-down tests periodically).
- **Upstream Filter Performance:** Monitor turbidity, ΔP, and backwash frequency of media filters or UF system feeding the RO.
- **ORP:** Check ORP reading before membranes to confirm dechlorination effectiveness (target $< 150\text{-}200$ mV).

10.4.4 Visual Inspections (Leaks, Noise, Vibration):

Walk around the skid regularly. Look for leaks from pipe joints, valve stems, pump seals, vessel end caps. Listen for unusual noises from pumps or motors (e.g., cavitation, bearing noise). Check for excessive vibration. Address any issues promptly.

10.4.5 Making Adjustments (Responding to Changes):

- **Temperature Changes:** As feed temperature changes, permeate flow will naturally fluctuate (colder = lower flow, warmer = higher flow). The control system (if automated with VFD) may adjust pump speed to maintain constant permeate flow, or the operator may need to adjust feed pressure manually. Do not mistake temperature-induced flow changes for fouling. This is why normalization is vital.
- **Feed Salinity Changes:** Variations in feed TDS will change osmotic pressure, affecting NDP and thus flux and rejection. Adjustments to feed pressure may be needed.
- **Recovery Adjustments:** Minor adjustments to the concentrate valve may be needed to maintain the target recovery rate as flows fluctuate, but never exceed the maximum recovery limit determined by scaling potential in the design.

Practical Insight: Develop a consistent routine. Don't just record numbers; look at them! Does anything seem unusual compared to the last reading or the baseline? Early detection is key to preventing minor issues from becoming major problems. Communicate any anomalies clearly between shifts.

10.5 Monitoring System Performance: Watching for Trends

Simply recording data isn't enough. Analyzing trends over time, particularly using normalized data, is crucial for effective performance management and predictive maintenance.

10.5.1 The Power of Normalization (Flux, Salt Passage, ΔP):

Why Normalize? As mentioned, raw operating data is heavily influenced by feed temperature, pressure, and salinity fluctuations. Normalization removes these external influences, revealing the true underlying performance changes caused by membrane fouling, scaling, or degradation.

What to Normalize:

- **Normalized Permeate Flux (or Flow):** Adjusts the measured flow to what it would be under reference conditions (e.g., baseline temperature, pressure, and feed concentration). A declining trend indicates loss of membrane permeability due to fouling/scaling.
- **Normalized Salt Passage (NSP) (or Rejection):** Adjusts the measured salt passage (permeate concentration / average feed concentration) to reference conditions. An increasing trend indicates worsening rejection, potentially due to fouling, scaling affecting membrane integrity, O-ring leaks, or chemical attack.
- **Normalized Differential Pressure (ΔP):** Adjusts the measured pressure drop across stages or vessels to reference flow conditions. An increasing trend clearly indicates plugging of feed channels by particulates, biofilm, or scale.

How: Performed using standard equations (often based on ASTM Standard D4516) that incorporate current flows, pressures, temperatures, and concentrations, and relate them back to the baseline reference conditions. Design software often includes normalization tools, or specialized spreadsheets/software can be used.

10.5.2 Trend Analysis: Detecting Fouling Early

Plot the Data: Regularly plot the normalized parameters (Flux, NSP, ΔP) versus time (days/weeks/months).

Look for Deviations: Compare current trends to the initial baseline and previous operating cycles.

Identify Fouling Signatures:

- Gradual decline in normalized flux + gradual increase in normalized ΔP often suggests particulate, colloidal, or organic fouling.
- Rapid increase in normalized ΔP, sometimes accompanied by flux decline, is characteristic of biofouling.
- Increase in normalized salt passage could indicate scaling disrupting the membrane, O-ring failure, or chemical damage.

Trigger Cleaning: Establish control limits (e.g., 10-15% decline in normalized flux, 15-20% increase in normalized ΔP, or 10-15% increase in NSP) to trigger membrane cleaning before fouling becomes severe and difficult to remove. Waiting too long makes cleaning less effective and risks permanent performance loss.

10.5.3 Key Performance Indicators (KPIs):

Beyond normalized data, track other KPIs like:

- Specific Energy Consumption (SEC - kWh per m^3 or per 1000 gallons of permeate): Tracks energy efficiency.
- Chemical Consumption Rates (antiscalant, SBS, cleaning chemicals): Tracks operating costs.
- Membrane Cleaning Frequency: Indicates overall system health and pretreatment effectiveness.
- Membrane Replacement Rate: Long-term indicator of system longevity.

Practical Insight: Data normalization and trend analysis transform raw numbers into actionable intelligence. This is the most powerful tool operators have for proactive maintenance and optimizing performance. Invest time in setting up and regularly reviewing these trends.

10.6 Shutdown Procedures: Protecting the System at Rest

How an RO system is shut down is just as important as how it's started, especially for protecting the membranes from scaling, biological growth, and degradation during idle periods.

10.6.1 Short-Term Shutdowns (Hours to Days)

10.6.1.1 The Critical Low-Pressure Permeate Flush:

This is the single most important shutdown step. Before stopping the feed pump completely, the system should undergo a flush cycle using low-pressure permeate water (or sometimes pretreated feed water if permeate isn't available, though less ideal).

Purpose: To displace the highly concentrated feed/concentrate water trapped within the membrane elements and feed channels. Leaving concentrated water stagnant invites scaling (as temperature changes affect solubility) and provides a rich environment for microbial growth.

Procedure: Typically automated. Feed pump ramps down, feed valve closes, concentrate valve opens fully, flush water supply valve opens, and flush water flows through the system at low pressure (e.g., 30-50 psi) for a set duration (e.g., 5-15 minutes) to effectively displace the concentrate. Flush water exits via the concentrate and sometimes permeate lines to drain.

10.6.1.2 Valve Positioning:

After the flush, the control system typically closes feed, permeate, and concentrate valves to keep the membranes wetted with low-TDS flush water.

10.6.2 Long-Term Shutdowns (Weeks to Months)

Simple flushing is insufficient for extended shutdowns, as residual microbes can still proliferate, and hydrolysis or oxidation can occur over time. Preservation is required.

10.6.2.1 Thorough Cleaning Before Shutdown:

It's highly recommended to perform a thorough chemical cleaning of the membranes before a planned long-term shutdown. This removes existing foulants and biofilm, ensuring the preservative solution can reach the membrane surface effectively.

10.6.2.2 Membrane Preservation Procedures:

After cleaning and flushing, the system is filled and soaked with a chemical preservative solution.

- **Common Preservative:** A solution of Sodium Bisulfite (SBS), typically 1% concentration by weight, dissolved in permeate water and adjusted to a slightly acidic pH (e.g., pH 4-5) is widely used. SBS acts as an oxygen scavenger and biostat (inhibits microbial growth).
- **Procedure:** The preservative solution is circulated through the RO system, ensuring all elements are filled. Valves are then closed to keep the membranes submerged in the solution.
- **Monitoring:** During storage, the preservative solution may need periodic checking (e.g., for pH and SBS concentration) and replenishment if necessary. The system should be protected from freezing temperatures.
- **Other Preservatives:** Formaldehyde or specialized commercial biocides were used historically but are less common now due to health, safety, and disposal concerns. Always follow membrane manufacturer recommendations for approved preservatives and procedures.

10.6.2.3 Protecting Equipment:

Drain water from non-membrane components (filters, pumps) where freezing or corrosion might occur. Follow manufacturer guidelines for preserving pumps and motors if shutdown is very long.

Practical Insight: Never skip the low-pressure permeate flush before any shutdown longer than a few hours. For TFC membranes, ensuring they remain wet at all times is critical. Preservation for long shutdowns is essential to avoid costly membrane replacement upon restart.

Restarting after preservation requires a thorough flush-out procedure similar to initial commissioning.

10.7 Chapter Summary and Key Takeaways

- Proper operation is crucial for RO system longevity, efficiency, and water quality.
- Pre-start checks (mechanical, electrical, pretreatment, valves, flushing) are essential to prevent damage.
- Initial start-up requires careful flushing, gradual pressurization, establishing baselines, and initial permeate diversion.
- Routine operation involves systematic monitoring and logging of key parameters (flows, pressures, quality, pretreatment checks, visual inspections).
- Performance normalization (adjusting flux, salt passage, ΔP for operating conditions) and trend analysis are vital for detecting fouling early and scheduling proactive maintenance.
- Shutdown procedures must protect membranes:
    - **Short-term:** Always perform a low-pressure permeate flush to displace concentrate.
    - **Long-term:** Clean membranes first, then fill and store with a preservative solution (e.g., 1% SBS).
- Adhering to recommended operating procedures, diligent monitoring, and proactive maintenance based on performance trends are the keys to successful, long-term RO operation.

10.8 References

1. AWWA Manual M46: Reverse Osmosis and Nanofiltration (2007). American Water Works Association. (Chapters on Operation, Monitoring, and Control).
2. DuPont Water Solutions, Hydranautics, Toray, LG Chem, etc. (Manufacturer Operation and Maintenance Manuals and Technical Bulletins): Provide specific procedures for start-up, shutdown, flushing, preservation, and performance monitoring relevant to their products. These are essential reading for operators.

3. ASTM Standard D4516-00(2014). Standard Practice for Standardizing Reverse Osmosis Performance Data. ASTM International. (Defines the methodology for data normalization).
4. Darton, E. G., & Fazel, M. (2001). Membrane autopsy: A valuable tool for RO system optimization. Desalination, 139(1-3), 225-233. (While focusing on autopsy, highlights operational issues leading to failure).
5. Wilf, M. (2007). The Guidebook to Membrane Desalination Technology: Reverse Osmosis, Nanofiltration and Hybrid Systems. Balaban Desalination Publications. (Contains practical operating insights).
6. Voutchkov, N. (2018). Desalination Engineering: Planning and Design. McGraw-Hill Education. (Includes sections on commissioning, operation, and monitoring).

Chapter 11: Membrane Fouling and Cleaning - Battling Performance Decline

11.1 Introduction: The Unavoidable Reality of Fouling

In an ideal world, the pretreatment system would remove every potential foulant, and the RO membranes would operate indefinitely with stable performance. In reality, even the best pretreatment systems aren't perfect, and operational fluctuations can occur. Over time, constituents from the feed water inevitably accumulate on the membrane surfaces or within the feed channels. This accumulation, collectively known as fouling, hinders membrane performance, leading to decreased water production (lower flux), increased energy consumption (higher feed pressure needed), higher operating pressure drop, and potentially poorer water quality (lower rejection).

While effective pretreatment (Chapter 9) and diligent operation (Chapter 10) are crucial for minimizing the rate of fouling, they cannot eliminate it entirely. Therefore, periodic chemical cleaning of the RO membranes is a necessary and routine part of operating almost any RO system. This chapter delves into the common types of fouling encountered, reinforces prevention strategies, discusses how to recognize the need for cleaning, and provides a detailed guide to the chemical cleaning process itself – selecting the right chemicals, applying them correctly, and ensuring safety and effectiveness. Proper cleaning restores membrane performance, extends membrane life, and maintains the overall efficiency of the RO plant.

11.2 Know Your Enemy: Types of Fouling

Understanding the nature of the foulant is the first step toward effective cleaning. Fouling is rarely composed of a single substance but is often a complex mixture. However, we can categorize the primary types based on their composition and origin:

11.2.1 Scaling (Inorganic Fouling): Mineral Precipitation

What: Deposition of inorganic mineral salts that have exceeded their solubility limit in the concentrated water near the membrane surface.

Common Types: (Recap from Ch 3 & 8)

- **Calcium Carbonate ($CaCO_3$):** Most common, whitish scale, especially in BWRO. Strongly pH-dependent. Predicted by LSI.
- **Calcium Sulfate ($CaSO_4$):** Gypsum. Less common but harder scale, solubility less pH-dependent.
- **Silica (SiO_2):** Can form amorphous or polymeric scale, especially at higher pH and concentration. Can be difficult to remove.
- **Barium Sulfate ($BaSO_4$) & Strontium Sulfate ($SrSO_4$):** Very insoluble, form hard scale even from low feed concentrations at high recovery. Extremely difficult to remove.
- **Metal Oxides/Hydroxides:** Iron ($Fe(OH)_3$ - reddish/brown), Manganese ($Mn(OH)_x$ - blackish), Aluminum ($Al(OH)_3$ - gelatinous white). Often linked to inadequate pretreatment (e.g., oxidation of dissolved metals, coagulant carryover).

Appearance: Often crystalline, hard deposits, though metal hydroxides can be more amorphous or gelatinous.

Primary Cause: Exceeding solubility limits due to concentration (high recovery/Beta) and Concentration Polarization (CP). Influenced by pH, temperature, and presence of other ions.

Primary Indicator: Often, a gradual decline in normalized flux. Sometimes an increase in salt passage if scale physically damages the membrane or disrupts flow. Can also contribute to increased ΔP.

11.2.2 Biofouling (Microbial Fouling): The Slime Menace

What: Growth of microorganisms (bacteria, fungi, algae) on membrane surfaces and feed spacers, embedded within a protective slime layer of Extracellular Polymeric Substances (EPS).

Mechanism: Attachment -> Colonization -> Growth -> EPS Production -> Biofilm Maturation.

Appearance: Slimy, gelatinous coating, often greenish, brownish, or grayish depending on the organisms involved. Can have characteristic odors (musty, swampy).

Primary Cause: Presence of viable microbes and sufficient nutrients (like biodegradable organics) in the feed water, especially under favorable temperature conditions and stagnant periods.

Primary Indicator: Often a rapid increase in differential pressure (ΔP) across stages, sometimes exceeding the rate of flux decline. Normalized flux will also decrease significantly. Can sometimes lead to increased salt passage.

11.2.3 Organic Fouling: Sticky Molecules

What: Adsorption or deposition of dissolved or colloidal organic molecules onto the membrane surface.

Common Types: Natural Organic Matter (NOM - humic/fulvic acids, common in surface water), proteins, oils, greases, antiscalant polymers (if overdosed or incompatible), synthetic organic chemicals.

Mechanism: Hydrophobic interactions, electrostatic attraction, or chemical binding between organic molecules and the membrane polymer (especially polyamides).

Appearance: Can range from a thin, invisible adsorbed layer to a thicker, often brownish or yellowish coating. Oils/greases form a distinct greasy film.

Primary Cause: Presence of specific organic foulants in the feed water that have an affinity for the membrane surface. Can be exacerbated by interactions with other foulants (e.g., bridging by calcium ions).

Primary Indicator: Typically causes a gradual decline in normalized flux. May contribute modestly to ΔP increase. Often makes subsequent biofouling worse by providing a conditioned surface and nutrients.

11.2.4 Colloidal Fouling: Fine Particle Accumulation

What: Deposition of extremely fine inorganic or organic particulate matter that passed through the pretreatment system.

Common Types: Clays, silt, silica colloids, fine metal oxides (iron, aluminum), bacterial fragments.

Mechanism: Particles accumulate on the membrane surface and within feed spacer channels due to hydrodynamic forces and surface interactions. Can be worsened by coagulant carryover or ineffective upstream filtration.

Appearance: Often forms a dense, paste-like layer. Color depends on the colloid type (e.g., clays are often tan/gray, iron is reddish-brown).

Primary Cause: Inadequate pretreatment failing to remove particles in the sub-micron to few-micron range (indicated by high SDI).

Primary Indicator: Primarily causes an increase in differential pressure (ΔP) due to feed channel plugging. Also contributes to flux decline.

11.2.5 Combinations and Interactions:

It's crucial to remember that fouling is rarely one pure type. Biofilms often trap colloidal particles. Organic layers can provide nutrients for microbes. Scaling can occur underneath or within other fouling layers. This complexity often necessitates multi-step cleaning procedures using different types of chemicals.

Table 11.1: Common RO Fouling Types and Indicators

Fouling Type	Primary Cause(s)	Key Indicator(s)	Common Appearance
Scaling (Inorganic)	Exceeding mineral solubility limits (high recovery, CP)	Flux decline, sometimes increased salt passage, ΔP increase	Crystalline deposits, hard layer
Biofouling (Microbial)	Microbial growth, EPS production	**Rapid ΔP increase**, flux decline	Slimy, gelatinous layer
Organic	Adsorption of NOM, oils, polymers, etc.	**Flux decline**, may increase ΔP	Thin films, brownish/yellowish
Colloidal	Inadequate removal of fine particles (high SDI)	**ΔP increase**, flux decline	Dense, paste-like layer

11.3 Prevention is Better Than Cure: Fouling Prevention Strategies Recap

Before diving into cleaning, it's worth briefly reiterating the primary defense lines covered in previous chapters:

- **Effective Pretreatment (Ch 8 & 9):** The most critical step. Tailoring coagulation, filtration (media/MF/UF), activated carbon, etc., to the specific raw water quality to meet RO feed guidelines (low SDI, low TOC, low microbial load).
- **Correct Chemical Conditioning (Ch 9):** Proper selection and dosing of antiscalants, biocides (if used), and dechlorinating agents (SBS). Accurate pH control if used.
- **Optimized RO Design & Operation (Ch 7 & 10):** Operating within recommended flux limits, maintaining adequate cross-flow velocity (appropriate staging/recycle), avoiding excessive recovery rates, and performing regular low-pressure permeate flushes before shutdowns.

11.4 Recognizing the Symptoms: Detecting Fouling Onset

As emphasized in Chapter 10, proactive monitoring using normalized performance data is key to detecting fouling early, when it's easiest to remove. The primary indicators are:

- **Decrease in Normalized Permeate Flux (or Flow):** Indicates loss of membrane permeability.
- **Increase in Normalized Differential Pressure (ΔP):** Indicates plugging of feed channels.
- **Increase in Normalized Salt Passage (NSP):** Indicates compromised rejection.

11.5 Restoring Performance: Chemical Cleaning of RO Membranes

When performance monitoring indicates that fouling has occurred and reached a pre-determined action limit, chemical cleaning (Clean-In-Place or CIP) is necessary.

11.5.1 When to Clean: Acting Proactively

Don't wait until performance has severely degraded! Fouling becomes much harder to remove, and the risk of irreversible damage increases, if cleaning is delayed. General industry guidelines trigger cleaning when:

- Normalized permeate flow decreases by 10-15% from the baseline.
- Normalized salt passage increases by 10-15% from the baseline.
- Normalized differential pressure (per element or per stage) increases by 15-20% from the baseline.

Adhering to these triggers helps maintain optimal performance and extend membrane life. Cleaning frequency can vary widely (from weeks to many months) depending on feed water quality, pretreatment effectiveness, and operating conditions.

11.5.2 The Cleaning Toolkit: Common Cleaning Chemicals

Different foulants require different chemical approaches. Most RO cleaning programs involve using specific low pH (acidic) and/or high pH (alkaline) cleaners, sometimes with additives, targeted at the suspected foulant(s). Always consult the membrane manufacturer's guidelines for compatible chemicals, concentration ranges, pH limits, and temperature limits. Using incompatible chemicals can severely damage the membranes.

11.5.2.1 Acids (Low pH Cleaners): Targeting Scale

Target: Primarily inorganic scales, especially calcium carbonate ($CaCO_3$) and metal oxides/hydroxides (Iron, Aluminum).

Mechanism: Dissolve acid-soluble scales by lowering the pH.

Common Examples:

- **Citric Acid:** Organic acid, relatively safe, good chelating properties for iron. Often used at pH ~3-4.
- **Hydrochloric Acid (HCl):** Strong mineral acid, very effective for $CaCO_3$, but highly corrosive and must be used with extreme caution and at low concentrations (e.g., pH not below 2.0, sometimes higher limits apply). Requires compatible materials of construction. Check manufacturer limits strictly.
- **Other Acids:** Phosphoric acid, specialty acid blends. Sulfuric acid is generally not recommended due to risk of precipitating $CaSO_4$.

Considerations: Use inhibited acids if available to minimize corrosion. Ensure pH does not drop below membrane manufacturer's absolute minimum limit (typically pH 2.0 for TFC, higher for some specialty types).

11.5.2.2 Alkaline Cleaners (High pH Cleaners): Tackling Organics, Biofilm, Silica

Target: Organic fouling (NOM, oils), biofilms (EPS matrix), and, to some extent, silica scale (solubility increases at high pH).

Mechanism: High pH helps hydrolyze and solubilize organic matter and EPS. Detergents often included enhance lifting/dispersion. High pH increases silica solubility.

Common Examples:

- **Sodium Hydroxide (NaOH - Caustic Soda):** Base chemical used to raise pH, typically to pH 11-12 (check membrane limit).
- **Proprietary Alkaline Cleaners:** Often NaOH-based but contain added chelants, detergents/surfactants for enhanced performance. Recommended by membrane manufacturers (e.g., DOW™ FilmTec™ RO Cleaners, Hydranautics RO Chemicals, Avista Cleaners).

Considerations: High pH cleaning is often significantly more effective at elevated temperatures (e.g., 35-40°C), up to the membrane's temperature limit. Handling caustic requires significant safety precautions. Thorough rinsing is needed to remove residual cleaner.

11.5.2.3 Chelating Agents: Enhancing Scale Removal

Target: Metal oxides, calcium sulfate, barium/strontium sulfate, silica (partially).

Mechanism: Form stable, water-soluble complexes (chelates) with metal ions, effectively dissolving otherwise insoluble scales.

Common Examples:

- **EDTA (Ethylenediaminetetraacetic Acid):** Common chelating agent, often added to alkaline cleaners. Effectiveness is pH-dependent.

- **Proprietary Blends:** Many commercial cleaners contain specific chelants tailored for RO applications.

Considerations: Help tackle scales not readily removed by simple acid or base cleaning.

11.5.2.4 Detergents/Surfactants: Lifting Oils and Biofilms

Target: Organic foulants (especially oils/greases), biofilms.

Mechanism: Reduce surface tension, emulsify oils, help penetrate and lift deposits from the membrane surface.

Common Examples: Anionic, non-ionic, or amphoteric surfactants. Often included in formulated alkaline cleaners. Check manufacturer compatibility.

Considerations: Need to be thoroughly rinsed out, as residual surfactant can cause foaming.

11.5.2.5 Biocides (Non-oxidizing): Killing Microbes

Target: Specifically killing bacteria, fungi, algae in biofilms.

Mechanism: Various modes of action depending on the chemical (e.g., disrupting cell walls, interfering with metabolism). Crucially, these must be non-oxidizing to avoid damaging TFC membranes.

Common Examples: DBNPA (2,2-dibromo-3-nitrilopropionamide), Isothiazolones, Glutaraldehyde (use declining). Always check membrane manufacturer approval.

Considerations: Used when biofouling is confirmed or strongly suspected. Often applied as a separate cleaning step or sometimes rotated with high pH cleaning. Requires specific contact times. May require special handling and disposal procedures. Never use oxidizing biocides like chlorine, peroxide, or ozone directly on TFC membranes.

Table 11.2: Matching Cleaning Agents to Fouling Types (General Guide)

Foulant Type	Primary Cleaning Approach	Common Chemicals / Conditions
Calcium Carbonate Scale	**Low pH (Acid)**	Citric Acid (pH ~3-4), HCl (pH >2.0, use cautiously)
Metal Oxides/Hydroxides	**Low pH (Acid)** + Chelant	Citric Acid, EDTA (often in alkaline clean first)
Calcium/Barium/Sr Sulfate	High pH + **Chelant**	Alkaline cleaner with EDTA, specific vendor products
Silica Scale	**High pH (Alkaline)** + Temp	NaOH (pH 11-12), high temperature (e.g., 35-40°C)
Organics (NOM, oils)	**High pH (Alkaline)** + Detergent	NaOH + Detergent, proprietary alkaline cleaners
Colloids (Silt, Clay)	**High pH (Alkaline)** + Detergent	NaOH + Detergent, sometimes low pH if metal-based
Biofouling (Biofilm)	**High pH (Alkaline)** + Detergent	NaOH + Detergent (targets EPS slime)
Biofouling (Microbial Kill)	**Non-oxidizing Biocide**	DBNPA, Isothiazolones (after or alternating w/ high pH)

Note: This is a general guide. Optimal strategy often involves sequential cleaning (e.g., high pH then low pH, or vice versa) and depends on specific foulant mixture.

11.5.3 Choosing the Right Weapon: Selecting Cleaning Agents

The most effective cleaning strategy depends on accurately identifying the dominant foulant(s). This relies on:

- **Reviewing Operating Data:** Trends in flux, ΔP, and NSP provide clues (see Table 11.1).

- **Feed Water Analysis:** Knowing potential scalants and foulants in the source water.
- **Pretreatment Performance:** Upsets upstream can point to likely foulants (e.g., coagulant carryover, high SDI).
- **Previous Cleaning History:** What worked well (or didn't) in the past?
- **Visual Inspection:** Examining cartridge filters or membranes during replacement can provide visual clues (color, texture).
- **Membrane Autopsy (Chapter 12):** The definitive method for identifying foulants when performance cannot be restored or problems persist.
- **Manufacturer Recommendations:** Consult membrane and chemical vendor guidelines. They often provide decision trees or recommendations based on symptoms.
- **Sequential Cleaning:** Often, a two-step cleaning using both high pH (alkaline) and low pH (acid) cleaners is recommended, especially if the foulant composition is mixed or uncertain. The order depends on the suspected primary foulant (e.g., tackle scale first with acid if LSI was high, or tackle organics/biofilm first with high pH if ΔP increased rapidly). A thorough flush is required between steps.

11.5.4 Optimizing the Attack: Key Cleaning Parameters

Using the right chemical isn't enough; it must be applied correctly.

11.5.4.1 Temperature:

Most chemical reactions proceed faster at higher temperatures. Cleaning effectiveness generally increases significantly with temperature, up to the membrane's maximum allowable cleaning temperature (typically 35-45°C for TFC, check specs). Using heated cleaning solutions (via CIP tank heater) is highly recommended, especially for alkaline cleaning of organics, biofilms, and silica.

11.5.4.2 pH:

Maintain the pH within the target range recommended for the specific chemical and foulant throughout the cleaning cycle. Monitor pH during circulation, as reactions can consume acid or base, and add more chemical as needed to maintain the target. Stay within the membrane's allowable pH limits (e.g., pH 2-12 for typical TFC cleaning ranges).

11.5.4.3 Flow and Pressure:

Circulate the cleaning solution at a moderate flow rate, typically similar to or slightly lower than the normal operating feed flow per vessel. The goal is good distribution and mixing, not high pressure. Keep pressures low (e.g., < 60 psi) to avoid compacting the foulant layer and to maximize chemical contact rather than forcing permeation. Ensure cross-flow velocity is sufficient to sweep away dislodged foulants.

11.5.4.4 Time (Soaking and Circulation):

Typical circulation time per cleaning step is 60-90 minutes. For stubborn foulants (heavy biofilm, certain scales), an additional soaking period (e.g., 1-8 hours, sometimes overnight, with pump off) may be recommended after initial circulation to allow chemicals more time to penetrate and react. Follow vendor guidance.

11.5.4.5 Concentration:

Use the chemical concentration recommended by the manufacturer. Higher concentrations are not always better and can sometimes damage membranes or be harder to rinse. Lower concentrations may be ineffective.

11.5.5 The Cleaning Procedure:

Step-by-Step Execution: Using a dedicated CIP skid (Chapter 6) is standard practice. A typical cleaning cycle involves:

1. Preparation:

- Identify target foulant(s) and select appropriate cleaning chemical(s) and sequence.
- Calculate required chemical quantities based on system volume (vessels, piping, CIP tank) and target concentration.

- Prepare a cleaning solution in the CIP tank using good quality water (RO permeate preferred, or dechlorinated soft water) to avoid introducing foulants/scalants.
- Heat solution to target temperature (if applicable). Adjust initial pH if necessary.

2. System Preparation:

- Perform a normal RO system shutdown, including the low-pressure permeate flush.
- Isolate the RO skid section to be cleaned from feed supply and product water tanks.
- Connect CIP skid supply and return hoses to the designated cleaning ports on the RO skid manifolding. Ensure correct flow direction (same as normal feed flow).

3. Initial Flush (Optional but Recommended):

Briefly flush the section with permeate from the CIP tank to displace any remaining feed water before introducing chemicals.

4. **Introduce**

Cleaning Solution: Start the CIP pump and slowly introduce the cleaning solution, displacing the flush water to drain until the solution starts returning to the CIP tank (check visually or via conductivity/pH).

5. Recirculate:

Close the drain path and recirculate the cleaning solution through the RO membranes and back to the CIP tank at the recommended flow rate and low pressure for the specified time (e.g., 60 mins). Monitor temperature, pH, pressure, and solution color/turbidity (darkening/clouding indicates foulant removal). Adjust pH/temperature as needed.

6. Soak (Optional):

If recommended, stop the CIP pump and allow the membranes to soak in the cleaning solution for the specified duration. Recirculate briefly afterwards if suggested.

7. Flush Out Cleaner:

Drain the dirty cleaning solution from the system and CIP tank (following proper disposal procedures). Thoroughly flush the system with good quality water (permeate preferred) until the pH, conductivity, and temperature of the flush water exiting the system return to normal feed/permeate water levels. This step is critical to remove all residual cleaning chemicals before returning to service. Insufficient rinsing can damage membranes or contaminate product water.

8. Return to Service:

Disconnect CIP hoses, restore normal valve positions, and perform a normal RO system start-up, including an initial permeate dump to drain to ensure any trace residuals are removed before sending water to service.

9. Record Keeping:

Document the entire cleaning process: date, chemicals used, concentrations, temperatures, pH values, circulation/soak times, initial and final RO performance data. This helps track cleaning effectiveness and optimize future cleanings.

11.5.6 Safety First: Handling Cleaning Chemicals

- **PPE:** Always wear appropriate Personal Protective Equipment (PPE) when handling cleaning chemicals: chemical-resistant gloves, safety goggles/face shield, apron/chemical suit.
- **Ventilation:** Ensure adequate ventilation when mixing and circulating chemicals, especially acids and biocides.
- **Mixing Order:** Follow safe mixing procedures (e.g., always add acid/caustic slowly to water, not vice versa).

- **MSDS:** Have Material Safety Data Sheets (MSDS) readily available for all chemicals used and follow their specific handling recommendations.
- **Training:** Ensure personnel are properly trained in safe chemical handling and emergency procedures.

11.5.7 Environmental Considerations:

Cleaning Solution Disposal: Spent cleaning solutions may contain high/low pH, dissolved foulants (including metals), and residual chemicals. They must be disposed of in accordance with local environmental regulations. This often requires neutralization (adjusting pH to near neutral) and potentially other treatment before discharge to sewer or off-site disposal. Check permits and regulations.

11.5.8 Assessing Cleaning Effectiveness: Did it Work?

The success of a cleaning is measured by the restoration of normalized performance parameters:

- **Successful:** Normalized flux, ΔP, and salt passage return to (or close to) the baseline values recorded after the previous successful cleaning.
- **Partially Successful:** Some improvement, but performance not fully restored. May indicate very stubborn fouling, incorrect chemical selection/procedure, or underlying membrane damage.
- **Unsuccessful:** Little or no improvement. Requires further investigation (Chapter 12), potentially including membrane autopsy, to determine the cause and appropriate next steps. Repeat cleaning with different chemicals might be attempted, but consider the risk of membrane damage.

11.6 Chapter Summary and Key Takeaways

- Fouling of RO membranes by scaling, biofilm, organics, or colloids is inevitable over time.

- Understanding the type of fouling (based on data, analysis, history) is key to selecting the appropriate chemical cleaning strategy.
- Cleaning should be triggered proactively based on normalized performance decline (e.g., 10-15% change in flux, ΔP, or NSP) before fouling becomes severe.
- Common cleaning agents include acids (for scale), alkaline cleaners with detergents/chelants (for organics, biofilm, silica), and non-oxidizing biocides (for microbial kill). Never use oxidizers like chlorine on TFC membranes.
- Effective cleaning requires optimizing temperature, pH, flow rate, contact time, and chemical concentration, following membrane and chemical manufacturer guidelines.
- A typical CIP procedure involves preparation, flushing, chemical circulation, optional soaking, thorough rinsing, and performance verification.
- Safety precautions (PPE, ventilation) are paramount when handling cleaning chemicals.
- Proper disposal of spent cleaning solutions according to regulations is essential.
- Cleaning effectiveness is judged by the restoration of normalized performance. If ineffective, further troubleshooting is needed.

11.7 References

1. AWWA Manual M46: Reverse Osmosis and Nanofiltration (2007). American Water Works Association. (Chapter on Membrane Fouling and Cleaning).
2. DuPont Water Solutions, Hydranautics, Toray, LG Chem, etc. (Manufacturer Technical Manuals and Bulletins): Provide detailed cleaning procedures, chemical compatibility lists, temperature/pH limits, and recommended cleaning triggers for their specific membrane products. These are primary references.
3. Antiscalant and Cleaning Chemical Manufacturer Literature (Avista, Nalco, Kemira, Suez, etc.): Offer specific product

information, application guidelines, and safety data sheets (MSDS).

4. Ridgway, H. F., & Flemming, H. C. (1996). Biofouling of membranes. In Mallevialle, J., Odendaal, P. E., & Wiesner, M. R. (Eds.), Water Treatment Membrane Processes. McGraw-Hill.
5. Ebrahim, S., Abdel-Jawad, M., Bou-Hamad, S., & Safar, M. (1997). Fifteen years of R&D in membrane desalination at KISR Part II. Membrane cleaning and regeneration. Desalination, 114(2), 167-184. (Example of research on cleaning effectiveness).
6. Vrouwenvelder, J. S., van der Kooij, D. (2001). Diagnosis of fouling problems of NF and RO membranes. Water Science and Technology: Water Supply, 1(4), 17-24. (Discusses diagnosing fouling types).
7. Pervov, A. G. (1991). Scale formation prognosis and cleaning procedure schedules in reverse osmosis systems operation. Desalination, 83(1-3), 77-118. (Early work on predicting scaling and cleaning).

Chapter 12: Troubleshooting - When Things Go Wrong

12.1 Introduction: The Art and Science of Troubleshooting

Sooner or later, every RO system operator or technician encounters a situation where the system isn't behaving as expected. Alarms trigger, water quality declines, flows drop, or pressures rise unexpectedly. The temptation might be to jump to conclusions or try quick fixes, but effective troubleshooting demands a more methodical approach. It's like being a detective arriving at a crime scene: you need to gather evidence (data), identify suspects (potential causes), conduct interrogations (perform tests), and follow the clues logically to solve the case.

This chapter provides a framework for investigating RO system problems. We'll start with the crucial first step – assessing the available data – and then explore common symptoms and their likely causes. We'll look beyond the membranes to mechanical issues and delve into advanced diagnostic techniques like membrane autopsy and integrity testing. The goal is to equip you with a systematic process to diagnose problems accurately and implement effective solutions, minimizing downtime and restoring optimal performance. Remember, guesswork can be costly; methodical investigation saves time, money, and potentially, the membranes themselves.

12.2 The Foundation: Data Assessment and Trend Analysis

Your most powerful troubleshooting tools are the operating data collected routinely (Chapter 10) and the performance trends derived from that data. Before touching any equipment, stop and analyze the data.

12.2.1 Is the Data Reliable? (Calibration Check)

Decisions based on faulty data are dangerous. If readings seem illogical or inconsistent, question the instruments first.

- Are gauges functioning? Are transmitters calibrated? When was the last calibration?
- Compare redundant sensors if available. Do they agree?
- Visually inspect sensors – are they coated or damaged?
- **Never assume the data is correct.** If a critical instrument is suspect, verify its calibration before proceeding.

12.2.2 Normalized Data: Seeing Through the Noise

Reiteration is necessary: Analyzing raw data (flow, pressure, conductivity) without normalization can be highly misleading due to the strong influence of temperature, pressure, and feed salinity changes. Always use normalized performance data (normalized flux, normalized salt passage/rejection, normalized ΔP) calculated against a consistent baseline (usually initial startup or post-cleaning performance). This reveals the true changes in membrane performance due to fouling, scaling, or damage.

12.2.3 Trend Spotting: Gradual vs. Sudden Changes

How did the problem develop? Looking at trend charts is crucial:

- **Gradual Decline/Increase:** Typically suggests a progressive process like gradual fouling (particulate, organic), slow scaling build-up, or membrane compaction/hydrolysis over time.
- **Sudden Change:** Points to an event – e.g., a pretreatment upset (chlorine breakthrough, filter failure leading to high SDI), O-ring failure, pump failure, valve malfunction, incorrect chemical dosing, or possibly rapid biofouling onset.
- **Cyclical Changes:** Might correlate with feed water quality variations (e.g., seasonal changes, tidal influences), inconsistent chemical dosing, or perhaps issues related to cleaning cycles.

12.2.4 Looking Upstream: Pretreatment Clues

Many RO problems originate in the pretreatment system. Always examine the pretreatment operating logs and data for the period leading up to the RO issue:

- Did SDI increase recently? (Indicates particulate/colloidal issue).
- Did turbidity spike?
- Was the cartridge filter ΔP high or changing frequently?
- Was there a chlorine breakthrough (check ORP logs!)?
- Were chemical dosing pumps operating correctly? Did tanks run empty? Was the correct chemical used?
- Were there any upsets in upstream filters (e.g., media filter backwash problems)?
- Did the feed water source analysis change?

Practical Insight: Dedicate sufficient time to data analysis before making adjustments or scheduling actions like cleaning. Often, the data holds the key to the root cause. Don't just look at the RO skid data in isolation; consider the entire process train.

12.3 Common Symptoms and Potential Causes

Here's a breakdown of common RO system problems and a checklist of potential causes to investigate, using the data analysis as your guide:

12.3.1 Problem: Low Permeate Flow (Normalized Flux Decline)

- **Fouling/Scaling:** Most common cause. Indicated by declining normalized flux, often accompanied by increasing normalized ΔP or feed pressure requirement. Identify likely foulant type based on trends and source water (see Ch 11). -> Action: Schedule appropriate chemical cleaning.
- **Low Feed Pressure:** Is the feed pump discharge pressure lower than normal? Check pump performance, suction strainers, possible internal wear.

- **Low Feed Temperature:** Raw flow will be low, but normalized flow should be stable unless fouled. Verify temperature sensor accuracy.
- **High Feed TDS / Osmotic Pressure:** Has feed salinity increased? This reduces NDP, lowering flux. Check feed conductivity.
- **Excessive Permeate Backpressure:** Is pressure building up in the permeate line due to downstream restrictions or tank levels?
- **Membrane Compaction:** Long-term, irreversible loss of flux due to high pressure operation, especially with older CA membranes or operation above limits. Normalized flux declines slowly over years. -> Action: Consider membrane replacement if severe.
- **Incorrect Valve Position:** Is a feed or permeate valve partially closed?

12.3.2 Problem: High Permeate Flow (Uncommon)

- **Membrane Damage / Integrity Loss:** If accompanied by low rejection, suspect physical damage (e.g., chlorine attack, abrasion) or seal failure. -> Action: Investigate integrity (Section 12.6).
- **Lower Feed TDS / Osmotic Pressure:** Has feed salinity decreased significantly? This increases NDP, boosting flux. Check feed conductivity.
- **Higher Feed Temperature:** Raw flow increases. Check normalized flow to confirm if it's just temperature effect.
- **Instrument Error:** Is the flow meter reading accurately?

12.3.3 Problem: Low Salt Rejection / High Permeate Conductivity (Normalized Salt Passage Increase)

- **O Ring / Seal Failure:** Most common cause of sudden increase in permeate conductivity. Leaks occur at element interconnectors or end cap seals, allowing feed/concentrate to bypass the membrane directly into the permeate stream. -> Action: Requires integrity testing (Profiling/Probing) to locate the leaking vessel/element, followed by replacement of the faulty seal.

- **Membrane Chemical Attack (e.g., Chlorine):** Irreversible damage to the polyamide layer destroys rejection capability. Often affects lead elements first. Indicated by sudden or rapid increase in NSP, especially after a known or suspected oxidant exposure (check ORP!). -> Action: Requires membrane replacement. Identify and eliminate source of oxidant.
- **Membrane Fouling/Scaling (Severe):** Some types of fouling/scaling can physically damage the membrane surface or disrupt flow patterns enough to impair rejection. Often accompanied by high ΔP or low flux. -> Action: Attempt appropriate cleaning; if rejection not restored, suspect damage.
- **Membrane Compaction/Hydrolysis:** Gradual degradation over time can lead to slow increase in NSP.
- **Operating Too Close to Flux/Recovery Limits:** Excessive concentration polarization can increase salt passage.
- **pH Effects:** For weakly ionized species (like silica, boron), rejection is pH-dependent. Check feed pH stability.
- **Feed Temperature Increase:** Higher temperature slightly increases salt passage. Check normalized value.
- **Instrument Error:** Verify conductivity meter calibration.

12.3.4 Problem: High Differential Pressure (ΔP) (Normalized ΔP Increase)

- **Particulate/Colloidal Fouling:** Blockage of feed spacer channels. Classic symptom of high SDI feed or pretreatment upset. -> Action: Improve pretreatment. Clean membranes (alkaline cleaner often helps). Consider elements with wider feed spacers.
- **Biofouling:** Rapid ΔP increase is characteristic. Biofilm physically plugs channels. -> Action: Implement appropriate biocidal cleaning protocol. Review biofouling prevention strategies (flushing, sanitization).
- **Scaling:** Scale deposits can constrict channels. -> Action: Implement appropriate scale-targeted cleaning. Review antiscalant program and recovery limits.

- **Telescoping/Physical Damage:** If elements have shifted or been damaged during loading/operation, flow paths can be blocked.
- **Low Feed Flow / High Recovery:** Running at lower than design feed flow can cause velocity to drop too low, allowing solids to settle and increase ΔP. Pushing recovery too high concentrates foulants.
- **Instrument Error:** Are the pressure transmitters reading correctly and calibrated?

12.3.5 Problem: High Feed Pressure Requirement

- **Membrane Fouling/Scaling:** Reduced membrane permeability (low normalized flux) requires higher pressure to maintain target flow. Most common cause. -> Action: Clean membranes.
- **Increased Feed TDS / Osmotic Pressure:** Higher osmotic resistance requires higher feed pressure. Check feed conductivity.
- **Lower Feed Temperature:** Increased water viscosity requires higher pressure to maintain target flow. Check normalized flux to distinguish from fouling.
- **Low Pump Performance:** Pump worn or damaged, not delivering required head at target flow. Check the pump curve.
- **Downstream Restrictions:** High permeate back pressure requiring higher feed pressure.

12.4 Beyond the Membrane: Mechanical Evaluation

Don't automatically assume the membranes are the problem. Malfunctioning mechanical components can mimic membrane issues or be the root cause.

12.4.1 Pump Problems (Pressure, Flow, Noise, Vibration):

Low Discharge Pressure/Flow: Check suction strainer for blockage, check for cavitation (low NPSHa - check tank level, suction valve position), check for internal wear (impellers, wear rings), verify motor speed/VFD output, check pump rotation. Compare operating point to pump curve.

Noise/Vibration: Investigate cavitation, bearing failure, misalignment, loose mounting bolts, motor issues.

12.4.2 Valve Issues (Leaks, Position, Control):

Internal Leaks: Valves not sealing properly (e.g., concentrate valve leaking affecting recovery, flush valves leaking).

Incorrect Position: Manual valves left partially closed? Automatic valve failed to open/close (check actuator, solenoid, signal)?

Control Valve Problems: Modulating valve sticking, poor calibration, hunting (unstable control).

12.4.3 Instrumentation Faults (Inaccurate Readings):

As mentioned, faulty sensors (pressure, flow, conductivity, ORP, pH, temperature) can lead to incorrect diagnosis and improper control actions. Calibrate regularly!

12.4.4 Piping and Vessel Integrity (Leaks, Blockages):

External Leaks: Obvious water drips/sprays from joints, seals, valve packings.

Internal Leaks: Harder to spot (e.g., cracked pipe within a header).

Blockages: Debris lodged in pipes, valves, strainers. Check suction strainers regularly.

Vessel Issues: Leaking end cap seals, damaged vessel internals (rare).

12.5 When Cleaning Fails: Membrane Autopsy

If routine cleaning doesn't restore performance, or if chronic fouling persists despite operational adjustments, a membrane autopsy provides the most definitive diagnosis.

12.5.1 Justification: Why Autopsy?

- Identify unknown foulant(s) or confirm suspected ones.
- Determine the primary cause of irreversible performance loss.
- Assess physical or chemical damage to the membrane.
- Optimize cleaning strategies based on actual foulant composition.
- Provide evidence for warranty claims (if applicable).
- Guide adjustments to pretreatment or operating procedures.

12.5.2 The Process: Sampling, Shipping, Analysis

- **Element Selection:** Choose element(s) representative of the problem (e.g., lead element from fouled stage, element from vessel with poor permeate quality). Consult membrane manufacturers for guidance.
- **Careful Removal:** Remove the element(s) carefully to avoid further damage or disturbance of foulant layers.
- **Packaging:** Keep the element wet (wrap in plastic with permeate or preservative). Package securely according to shipper and laboratory instructions to prevent damage during transit.
- **Information:** Provide detailed operating history, water analyses, cleaning records, and description of the problem to the autopsy lab.
- **Analysis:** Performed by specialized laboratories (often membrane manufacturers or independent labs). Includes visual inspection, foulant probing/collection, weighing, integrity testing (offline), microscopic examination (SEM), elemental analysis (EDX/XRF), organic analysis (FTIR), microbiological analysis, and scale analysis (XRD).

12.5.3 What Autopsy Reveals: Definitive Foulant ID, Damage Assessment

The detailed report identifies the foulants present (inorganic scale types, metals, silica, organic functional groups, types of microbes), their location and distribution within the element, and assesses any physical damage (scratches, compaction, degradation) or chemical attack (e.g., oxidation). This provides concrete evidence to guide corrective actions.

Practical Insight: Autopsy is an investment (cost of element, shipping, analysis fee) but can save much more in the long run by preventing repeated cleaning failures or unnecessary replacement of all membranes. It turns guesswork into targeted action.

12.6 Finding the Leaks: Membrane and System Integrity Testing

When high permeate conductivity points towards a leak rather than general membrane failure, integrity testing helps locate the source without necessarily performing an autopsy immediately.

12.6.1 Locating the Source of Poor Quality: Profiling

Concept: Systematically measuring the conductivity of the permeate coming from each individual pressure vessel (requires sample ports on each vessel's permeate outlet).

Procedure: While the system is operating normally, collect permeate samples from each vessel in sequence and measure conductivity.

Interpretation: A vessel showing significantly higher conductivity than its neighbors likely contains a leaking element, interconnector O-ring, or end cap seal.

Action: Isolate the suspect vessel (if valving allows) or shut down the system and investigate the elements and seals within that specific vessel.

12.6.2 Pinpointing the Element: Probing

Concept: Taking permeate samples from different locations within a pressure vessel's central permeate tube. Requires special end caps with probe access ports.

Procedure: A thin tube is inserted down the permeate tube, and samples are drawn (using suction or system pressure) corresponding to the permeate produced by each element (or even sections within an element). Conductivity is measured for each sample.

Interpretation: Allows identification of the specific leaking element(s) within the vessel.

Action: Replace the identified leaking element or its seals.

12.6.3 Offline Gross Leak Detection:

Pressure/Vacuum Decay Tests

Concept: Used to detect significant leaks in individual elements or vessels when the system is offline.

Pressure Hold Test: Isolate the element/vessel, pressurize the feed side slightly with air (e.g., 15-30 psi), and monitor for pressure decay over time. Rapid decay indicates a leak.

Vacuum Decay Test: Apply a vacuum to the permeate side and monitor for vacuum loss.

Bubble Point Test: More common for MF/UF integrity, less sensitive for RO. Involves wetting the membrane and applying low air pressure to the feed side, looking for bubbles appearing on the permeate side at pressures lower than expected for intact membrane (indicating a pore/defect/leak).

Limitations: These tests usually only detect relatively large leaks, not subtle O-ring bypasses or minor membrane damage that might still impact high-purity applications.

Practical Insight: Profiling is a relatively simple and highly effective first step for locating permeate quality issues in multi-vessel systems, provided sample ports exist. Probing offers more precision but requires specialized hardware.

12.7 Other Diagnostic Tools and Tests

12.7.1 Cleaning Study (Off-line Membrane Sample):

If standard cleanings are ineffective, a small sample of fouled membrane (e.g., from an autopsied element or a deliberately fouled test coupon) can be sent to a lab. They perform bench-scale cleaning tests with various chemicals and conditions to determine the most effective cleaning recipe before attempting it on the full system.

12.7.2 Feed Water Analysis Verification:

If performance changes unexpectedly, especially permeate quality, get a new, comprehensive feed water analysis. Did the salinity, pH, or concentration of a specific foulant/scalant change significantly?

12.7.3 Detailed Pretreatment Evaluation:

Conduct a thorough audit of the entire pretreatment system – filter performance, chemical dosing accuracy, sensor calibrations – if it's suspected as the root cause.

12.8 The Systematic Approach: A Troubleshooting Workflow

1. **Define the Problem:** What specific symptom(s) are occurring? (Low flow, high conductivity, high ΔP?)
2. **Gather Data:** Collect current operating logs, historical trends (normalized!), pretreatment data, maintenance records.
3. **Verify Data:** Check instrument calibrations if readings are suspect.
4. **Analyze Trends:** Look for gradual vs. sudden changes. Correlate RO data with pretreatment data.

5. **Formulate Hypotheses:** Based on symptoms and data, list the most likely potential causes (fouling type, mechanical issue, leak, etc.).
6. **Prioritize & Test:** Start with the simplest, most likely, or easiest-to-check hypotheses.
   - Check valve positions, tank levels, basic parameters.
   - Verify pretreatment performance (SDI, Cl_2, dosing).
   - Perform mechanical checks (pumps, valves).
   - Perform integrity tests (profiling) if leaks suspected.
7. **Take Corrective Action:** Based on test results (e.g., perform targeted cleaning, replace seal, adjust chemical dose, repair pump).
8. **Monitor & Verify:** Confirm that the corrective action resolved the problem and performance returns to normal. Update baseline data if necessary (e.g., after successful cleaning).
9. **Document:** Record the problem, investigation steps, findings, corrective actions, and results. This builds a valuable knowledge base for future issues.
10. **Escalate if Necessary:** If the problem persists or the cause remains elusive, consider consulting the membrane manufacturer, chemical supplier, system vendor, or performing a membrane autopsy.

12.9 Chapter Summary and Key Takeaways

- Effective RO troubleshooting requires a systematic, data-driven approach.
- Start by assessing reliable operating data, especially normalized performance trends, and correlate with pretreatment data.
- Identify the specific symptom (low flow, low rejection, high ΔP, high feed pressure) and consider the most likely potential causes for that symptom.
- Investigate methodically, checking pretreatment, membrane performance indicators, mechanical components, and instrumentation.

- Use integrity testing (profiling, probing) to locate sources of poor permeate quality (leaks).
- If performance cannot be restored or the cause is unclear, a membrane autopsy provides definitive answers about fouling and membrane condition.
- Always consider safety during troubleshooting activities.
- Document the entire process to build knowledge and guide future actions.

12.10 References

1. AWWA Manual M46: Reverse Osmosis and Nanofiltration (2007). American Water Works Association. (Chapter on Troubleshooting).
2. DuPont Water Solutions, Hydranautics, Toray, LG Chem, etc. (Manufacturer Troubleshooting Guides and Technical Service Bulletins): Provide specific guidance related to their membranes and common problems. Essential resources.
3. Amjad, Z. (Ed.). (2011). The Science and Technology of Industrial Water Treatment. CRC Press. (Chapters cover fouling, cleaning, and troubleshooting aspects).
4. Chesters, S. P. (2003). Membrane autopsy–the best diagnostic tool available. Filtration & Separation, 40(8), 26-28.
5. Cohen, Y., & Semiat, R. (Eds.). (2017). Water-Energy Nexus: Fundamentals, Implementation, and Optimization of Water Purification and Desalination Systems. Wiley. (Includes advanced diagnostic discussions).
6. ASTM Standard D4516-00(2014). Standard Practice for Standardizing Reverse Osmosis Performance Data. (Basis for normalization needed for troubleshooting).
7. Membrane Autopsy Service Providers (e.g., manufacturers' labs, independent labs like Avista Technologies, Water Quality Engineers): Their websites and literature often describe the autopsy process and benefits.

Chapter 13: Industrial Applications of RO - Powering Progress

13.1 Introduction: RO as an Industrial Enabler

Modern industry relies heavily on water. It's used as a solvent, a coolant, a cleaning agent, an ingredient, a carrier for heat and power, and for countless other process needs. However, the quality of water required often far exceeds that available directly from municipal supplies or natural sources. Impurities like dissolved salts, minerals, suspended solids, organics, and microbes can cause corrosion, scaling, product contamination, process inefficiencies, and equipment failure.

Reverse Osmosis has emerged as a cornerstone technology for providing industries with water tailored to their specific, often demanding, quality requirements. Its ability to remove a broad spectrum of dissolved contaminants efficiently and without the use of regeneration chemicals (like ion exchange) makes it highly attractive. From producing the massive volumes of desalinated water needed by coastal refineries to generating the incredibly pure water required for microchip fabrication, RO plays a vital, often unseen, role in enabling modern industrial processes.

Furthermore, as environmental regulations tighten and water scarcity increases, RO is also becoming crucial for treating industrial wastewater, allowing industries to meet discharge limits, recover valuable water for reuse, and move towards more sustainable water management practices, including Zero Liquid Discharge (ZLD) goals. This chapter explores some of the most significant industrial sectors where RO technology is making a critical difference.

13.2 The Cornerstone Application: Desalination for Industry

While municipal desalination gets much attention, many large industrial facilities, particularly those located in coastal or arid regions, rely on desalination technologies, primarily RO, for their primary water supply.

13.2.1 Seawater RO (SWRO) for Coastal Industries:

- **Need:** Industries like oil refineries, petrochemical plants, power stations, and large manufacturing complexes located near coastlines often have enormous water demands that cannot be met by local freshwater resources. Seawater offers an abundant, drought-proof source.
- **Application:** Large-scale SWRO plants are built specifically to supply these industrial facilities with process water, cooling water makeup, boiler feedwater (after further polishing), and sometimes even potable water for on-site personnel.
- **Considerations:** These plants face the typical SWRO challenges: high salinity feed (35,000-45,000+ mg/L TDS), requiring high pressures (800-1200 psi) and significant energy consumption (though energy recovery devices mitigate this). Robust pretreatment is essential to handle variable seawater quality (algal blooms, turbidity events). Materials of construction must withstand high chloride levels.
- **Example:** Many refineries and LNG terminals in the Middle East operate dedicated, large-capacity SWRO plants.

13.2.2 Brackish Water RO (BWRO) for Inland Needs:

- **Need:** Inland industries located away from reliable surface water sources or in areas with poor quality municipal water may turn to brackish groundwater (1,000-15,000 mg/L TDS).
- **Application:** BWRO systems treat this groundwater to produce suitable quality water for various industrial uses. The required permeate quality dictates the system design (e.g., single pass for general use, double pass for high purity).

- **Considerations:** Lower operating pressures (150-400 psi) mean lower energy costs compared to SWRO. However, brackish waters often contain high concentrations of hardness (Calcium, Magnesium), bicarbonate alkalinity, silica, iron, manganese, or sometimes specific contaminants like barium or strontium, making scaling and fouling key concerns that require careful pretreatment design (antiscalants, pH adjustment, specific removal processes). Concentrate disposal can also be a significant challenge for inland facilities.
- **Example:** Mining operations in arid regions often use BWRO to treat saline mine water or brackish groundwater for process use.

13.3 Keeping the Lights On: Power Generation

The power industry is one of the largest industrial users of high-purity water. Maintaining water chemistry within strict limits is critical for preventing catastrophic failures in boilers, turbines, and cooling systems. RO is a central technology in power plant water treatment.

13.3.1 Boiler Feedwater: Preventing Scale and Corrosion

- **Need:** Water fed to boilers, especially high-pressure boilers used in thermal power plants, must be exceptionally pure. Even trace amounts of dissolved minerals (especially hardness ions like Ca^{2+}, Mg^{2+}, and silica SiO_2) can form insulating scale deposits on boiler tubes. Scale reduces heat transfer efficiency (wasting fuel) and can lead to localized overheating and tube rupture – a major safety hazard. Dissolved salts (like chlorides) and dissolved gases (like oxygen and carbon dioxide) contribute to corrosion.
- **RO's Role:** RO is highly effective at removing the vast majority (>98-99.5%) of dissolved ions, including hardness, silica, and chlorides, from the makeup water source (which could be municipal water, surface water, or desalinated water).
- **Typical Treatment Train:** Makeup water -> Pretreatment (filtration, etc.) -> Single or Double Pass RO -> Final

Polishing (Electrodeionization - EDI, or Mixed-Bed Ion Exchange - MB) -> Deaeration (to remove dissolved O_2/CO_2).

- **Benefits:** RO significantly reduces the ionic load on downstream polishing units (EDI/MB), extending their service life and reducing regeneration chemical usage and cost. It provides a reliable barrier against upsets in makeup water quality. Achieving very low silica levels (< 5-10 ppb) is often critical for high-pressure boilers to prevent silica volatilization and deposition on turbine blades, and RO plays a key role.

13.3.2 Cooling Tower Makeup and Blowdown Treatment

- **Need (Makeup):** Cooling towers evaporate large amounts of water to provide cooling. As water evaporates, dissolved minerals concentrate in the remaining circulating water. If makeup water has high hardness or TDS, this concentration effect (cycles of concentration) quickly leads to scaling and corrosion issues within the tower and heat exchangers. Using higher quality makeup water allows the tower to operate at higher cycles of concentration, reducing the amount of water discharged (blowdown).
- **RO's Role (Makeup):** RO can treat the makeup water source to remove hardness and dissolved solids, allowing higher cycles of concentration, saving water, and reducing chemical treatment costs (scale/corrosion inhibitors).
- **Need (Blowdown):** Cooling tower blowdown is the concentrated water intentionally drained from the system to control TDS levels. It can be high in salts, treatment chemicals, and sometimes contaminants scrubbed from the air, making direct discharge problematic.
- **RO's Role (Blowdown):** RO (often coupled with robust pretreatment to handle the complex chemistry) can be used to treat blowdown water, recovering a significant portion as clean water for reuse (e.g., back as makeup water) and producing a smaller, highly concentrated brine stream for disposal. This helps meet discharge regulations and achieve water conservation goals.

13.3.3 NOx Control Water (SCR/SNCR Systems)

- **Need:** Selective Catalytic Reduction (SCR) and Selective Non-Catalytic Reduction (SNCR) systems used to control Nitrogen Oxide (NOx) emissions from fossil fuel combustion (boilers, turbines, engines) often involve injecting ammonia or urea solutions into the flue gas. The water used to prepare these reagent solutions must be very pure to prevent plugging of injection nozzles and fouling/poisoning of the catalyst (in SCR systems).
- **RO's Role:** RO provides the necessary high-purity demineralized water, ensuring reliable operation of these critical environmental control systems.

13.4 The Foundation of Modern Tech: Electronics and Semiconductor Manufacturing

Perhaps no industry demands higher water purity than semiconductor manufacturing. The fabrication of microchips involves hundreds of intricate steps, many involving rinsing wafer surfaces. Even infinitesimal traces of ionic, organic, or particulate contamination in the rinse water can cause defects, ruining entire batches of valuable chips.

13.4.1 The Quest for Ultrapure Water (UPW)

Need: Water purity requirements are measured not in parts per million (ppm) or even parts per billion (ppb), but often in parts per trillion (ppt) for critical parameters like specific ions, total organic carbon (TOC), dissolved oxygen, silica, and particles. Resistivity, the inverse of conductivity, is used as a key measure, with target values often exceeding 18.18 Megohm-cm (the theoretical maximum for pure water at 25°C).

13.4.2 RO's Role in the Multi-Barrier UPW Train:

RO is an indispensable workhorse in the complex, multi-stage treatment train used to produce UPW. It's not the final step, but it performs the crucial bulk demineralization.

Typical UPW Train: Source Water -> Pretreatment (often extensive, including MF/UF) -> Double Pass RO -> UV Treatment (for TOC reduction and disinfection) -> Primary Ion Exchange (IX) or Electrodeionization (EDI) -> Polishing Mixed-Bed Ion Exchange -> Final Membrane Degasification (to remove dissolved O_2/CO_2) -> Sub-micron Filtration (e.g., 0.02 µm) at Point-of-Use.

RO's Function: The double-pass RO (often with inter-stage pH adjustment and sometimes membrane degassing) removes >99.9% of dissolved salts, silica, and larger organics. This dramatically reduces the load on the downstream polishing IX/EDI units, which handle the final trace ion removal. The reliability and efficiency of the RO system are critical to the overall performance and cost of the UPW plant.

13.4.3 Rinse Water Recycling:

The volume of UPW used for rinsing is enormous. Treating and recycling this "used" rinse water (which is still relatively clean compared to raw water) is essential for water conservation and cost reduction. RO is often employed in these reclaim systems to remove contaminants picked up during the rinsing process, allowing the water to be re-polished and reused.

13.5 Purity on the Plate: Food and Beverage Processing

Water quality is paramount in the food and beverage industry for product quality, consistency, safety, and regulatory compliance. RO finds diverse applications here.

13.5.1 Ingredient Water (Bottled Water, Soft Drinks, Brewing):

Need: The mineral content and taste of water directly impact the final product. Municipal water often contains chlorine, variable mineral levels, and potential off-tastes or odors unsuitable for beverages.

RO's Role: RO is widely used to produce highly purified, demineralized water ("RO water" or "purified water"). This provides a consistent, neutral base water ("blank slate") to which specific minerals or formulations can be added back precisely to achieve the desired taste profile for bottled water, soft drinks, juices, or beer. It also removes chlorine and other potential contaminants.

13.5.2 Boiler Feedwater and Utility Water:

Similar to the power industry, food/beverage plants use boilers for steam generation (for cooking, sterilization, heating). RO provides the high-purity feedwater needed to prevent scale and corrosion, ensuring efficient and safe boiler operation. RO water is also used for general utility purposes requiring low mineral content.

13.5.3 Product Concentration (Dairy Whey, Juices - often NF/RO combination):

Need: Removing water to concentrate products like fruit juices, milk, or whey (a byproduct of cheese making) increases shelf life, reduces shipping volume/cost, and produces valuable concentrates (like Whey Protein Concentrate - WPC).

RO/NF Role: While evaporation is traditional, membrane processes like RO and Nanofiltration (NF – a related process using slightly "looser" membranes that pass more monovalent ions but reject divalent ions and larger molecules) are increasingly used. They operate at lower temperatures (preserving sensitive flavors and nutrients) and can be more energy-efficient. RO removes water while retaining most solutes. NF might be used to selectively remove some salts (like sodium from whey) while retaining valuable proteins and lactose. Often used in combination.

13.5.4 Process Water (Washing, Rinsing, Cleaning):

Using RO water for washing fruits/vegetables, rinsing containers, or final rinses in Clean-In-Place (CIP) systems prevents spotting (mineral deposits) and ensures product purity.

13.5.5 Wastewater Treatment and Reuse:

Food processing generates wastewater often high in organic load (BOD), sugars, salts, and cleaning chemicals. RO (after appropriate biological and physical pretreatment) can treat this wastewater to recover high-quality water for reuse within the plant (e.g., for cleaning, cooling tower makeup), reducing freshwater intake and discharge volumes/costs.

13.6 Ensuring Safety and Efficacy: Pharmaceutical Production

The pharmaceutical industry operates under extremely strict regulatory requirements (e.g., US Pharmacopeia - USP, European Pharmacopoeia - EP) for water purity, as water is often a direct ingredient in final products or used for critical cleaning. RO is a fundamental technology in pharmaceutical water systems.

13.6.1 Meeting Stringent Standards (USP, EP):

Pharmacopoeias define specific grades of water, such as Purified Water (PW) and Water For Injection (WFI), with rigorous limits on conductivity, TOC, microbial counts (endotoxin levels for WFI), and specific ions.

13.6.2 Purified Water (PW) and Water For Injection (WFI) Generation:

- **RO's Role:** RO (often double-pass) is a core component in generating both PW and WFI. It provides the primary demineralization and removal of organics and microbes.

- **Typical Train (PW):** Municipal Water -> Pretreatment (Softening, Multimedia Filtration, Carbon Filtration) -> Single or Double Pass RO -> (Optional) EDI -> UV Disinfection -> Storage & Distribution Loop.
- **Typical Train (WFI):** PW -> Distillation (traditional method) or Membrane-based WFI generation (UF often following RO/EDI) -> Storage & Hot Distribution Loop (>65-80°C to prevent microbial growth). Note: Regulations increasingly accept membrane-based WFI systems.
- **Benefits:** RO provides a reliable barrier against ionic and microbial contaminants, reducing the load on downstream units and ensuring consistent feed quality for distillation or final polishing.

13.6.3 Sanitization and Validation Considerations:

Pharmaceutical water systems require frequent sanitization (using heat or chemicals like ozone - if compatible materials are used downstream of RO) to control microbial growth. RO systems in this service are often designed with sanitary connections, smooth surfaces, and materials (like 316L SS) suitable for sanitization. Hot water sanitizable RO membranes and modules are available.

The entire water system, including the RO unit, must undergo rigorous validation to prove it consistently produces water meeting the required specifications. This involves extensive testing, documentation, and qualification protocols (IQ/OQ/PQ).

13.7 Closing the Loop: Industrial Wastewater Treatment and Reuse

Treating industrial wastewater is a rapidly growing application for RO, driven by stricter environmental discharge regulations, rising freshwater costs, water scarcity, and corporate sustainability initiatives.

13.7.1 Treating Challenging Effluents:

Industrial wastewater streams can be highly variable and complex, containing salts, heavy metals, refractory organics, oils, suspended solids, and treatment chemicals. RO is often used as a tertiary treatment step after primary (physical) and secondary (biological/chemical) treatment have removed the bulk of suspended solids and organic load (BOD).

Robust Pretreatment is Key: Effective removal of suspended solids, colloids, organics, and potential scalants/foulants before the RO is absolutely critical for successful wastewater treatment. This often involves clarification, biological treatment (e.g., MBR - Membrane Bioreactor), MF/UF, and sometimes activated carbon.

13.7.2 Meeting Discharge Limits:

RO can remove dissolved salts (TDS) and specific contaminants (e.g., heavy metals, certain organics) to meet stringent discharge permit limits before the effluent is released to the environment.

13.7.3 Water Reuse and Zero Liquid Discharge (ZLD) Goals:

The high quality of RO permeate often makes it suitable for reuse within the industrial facility (e.g., cooling tower makeup, process water, boiler feed pre-treatment), significantly reducing freshwater consumption.

RO plays a key role in Minimal Liquid Discharge (MLD) and Zero Liquid Discharge (ZLD) systems. By concentrating the wastewater stream after initial treatments, RO minimizes the volume of the final brine that needs to be disposed of (often through expensive evaporators and crystallizers in ZLD systems).

13.7.4 Specific Industrial Examples:

- **Oil Refineries & Petrochemicals:** Treating sour water stripper effluent, cooling tower blowdown, and general process wastewater for reuse or discharge.
- **Mining:** Treating acid mine drainage (after neutralization/metals removal) or saline process water to recover water in arid regions.
- **Textile Industry:** Treating dyehouse effluent (often requiring extensive pretreatment for color and organics) to recover water and salts for reuse.
- **Pulp & Paper Industry:** Treating bleach plant effluent or general mill wastewater.
- **Landfill Leachate Treatment:** RO (with very robust pretreatment) is used to remove high concentrations of TDS, ammonia, and refractory organics.

Challenges in Wastewater RO: High fouling potential (biological, organic, scaling), feed variability, complex chemistry requiring sophisticated pretreatment and cleaning strategies, and concentrate management remain significant challenges.

13.8 Chapter Summary and Key Takeaways

- Reverse Osmosis is a versatile and critical technology enabling diverse industrial processes by providing tailored water quality.
- **Industrial Desalination (SWRO/BWRO):** Provides essential water supply for coastal and inland industries in water-scarce areas.
- **Power Generation:** RO produces high-purity boiler feedwater (preventing scale/corrosion), treats cooling tower makeup/blowdown, and provides water for NOx control.
- **Electronics/Semiconductors:** Double-pass RO is indispensable for generating Ultrapure Water (UPW) required for chip manufacturing and rinse water recycling.

- **Food & Beverage:** Used for ingredient water (consistency/purity), utility water, product concentration (with NF), and wastewater reuse.
- **Pharmaceuticals:** Core technology for producing USP Purified Water (PW) and feed for Water For Injection (WFI) systems, meeting stringent regulatory standards under validated conditions.
- **Industrial Wastewater Treatment & Reuse:** Increasingly vital for meeting discharge limits, recovering water for reuse, and achieving MLD/ZLD goals across various industries (refining, mining, textiles, etc.), though requiring robust pretreatment.
- The specific design, pretreatment, and operational considerations for RO vary significantly depending on the industrial application and its unique water quality requirements and challenges.

13.9 References

1. AWWA Manual M46: Reverse Osmosis and Nanofiltration (2007). American Water Works Association. (Overview chapter on applications).
2. Voutchkov, N. (2018). Desalination Engineering: Planning and Design. McGraw-Hill Education. (Covers industrial desalination and reuse).
3. Buros, O. K. (2000). The ABCs of Desalting (2nd ed.). International Desalination Association (IDA). (General overview including industrial uses).
4. Nalco Water. The Nalco Water Handbook (Multiple Editions). McGraw-Hill. (Industry handbook covering water treatment for power, cooling, boilers, etc., often including RO).
5. Meltzer, T. H., & Jornitz, M. W. (Eds.). (1998). High-Purity Water Preparation for the Semiconductor, Pharmaceutical, and Power Industries. Tall Oaks Publishing.
6. United States Pharmacopeia (USP). <Chapters on Water for Pharmaceutical Purposes>. (Defines standards for PW and WFI).
7. European Pharmacopoeia (Ph. Eur.). <Monographs on Water for Pharmaceutical Use>.

8. Frankiewicz, T. C., Dabir, B., & Gupta, S. (2005). Treating refinery wastewater using reverse osmosis. Environmental Progress, 24(2), 154-162. (Example application paper).
9. Mohsen, M. S., & Jaber, J. O. (2003). Potential of industrial wastewater reuse: a case study. Desalination, 158(1-3), 161-170.

Chapter 14: Specialized Applications - Meeting Critical Needs

14.1 Introduction: Tailoring RO for Specific Missions

While Chapter 13 showcased the widespread use of Reverse Osmosis across major industrial sectors, the technology's versatility extends to numerous specialized applications where the primary goal might be public health, scientific precision, agricultural viability, or meeting niche quality requirements. These applications often involve different scales of operation, unique contaminant challenges, specific regulatory frameworks, and sometimes post-treatment steps tailored to the end-use.

This chapter delves into some of these critical specialized roles for RO. We'll examine its use in providing safe drinking water at various scales, from large municipalities down to individual homes. We'll look at the production of ultrapure water specifically for laboratory and research settings. We'll explore the challenges and potential of using RO to provide suitable water for agriculture in arid regions. Finally, we'll revisit the pharmaceutical sector to highlight specific water uses beyond bulk PW and WFI generation. These applications underscore RO's adaptability in meeting some of society's most fundamental and demanding water needs.

14.2 The Most Fundamental Need: Potable Water Production

Ensuring access to safe, clean drinking water is a global public health priority. Reverse Osmosis plays an increasingly vital role in achieving this, particularly in areas facing water scarcity or contamination issues.

14.2.1 Large-Scale Municipal Desalination (SWRO & BWRO):

Context: This is perhaps the most visible application of RO. Coastal cities and arid regions worldwide facing dwindling freshwater supplies are turning to the sea or brackish groundwater as reliable sources.

Technology: Large desalination plants, often employing multiple high-capacity RO trains (as discussed in Chapter 5), produce millions of gallons (or cubic meters) of drinking water per day. SWRO treats seawater, while BWRO treats less saline sources like estuaries or inland aquifers.

Key Aspects: Focus is on reliability, energy efficiency (critical for SWRO, extensive use of Energy Recovery Devices - ERDs), robust pretreatment to handle variable source water, effective post-treatment (remineralization and disinfection), and environmentally sound concentrate management. These plants are major infrastructure projects requiring significant capital investment and sophisticated operation.

Examples: Numerous large plants in the Middle East (e.g., Sorek, Ashkelon in Israel; plants in UAE, Saudi Arabia), Australia (e.g., Perth, Sydney), USA (e.g., Carlsbad, Tampa Bay), Singapore, Spain, and many other locations rely heavily on RO for municipal supply.

14.2.2 Small Systems for Communities and Emergency Response:

Context: Smaller towns, remote communities, resorts, or disaster relief operations may also require desalination or treatment of contaminated sources, but at a much smaller scale.

Technology: Skid-mounted or containerized RO systems offer compact, transportable solutions. These can range from a few thousand GPD up to hundreds of thousands GPD. They often prioritize ease of operation, robustness, and sometimes energy independence (e.g., solar-powered units).

Applications: Providing drinking water for island communities, drought relief, military deployments, post-hurricane recovery efforts, or treating contaminated wells (e.g., high nitrates, arsenic).

14.2.3 Point-of-Use (POU) / Point-of-Entry (POE) Systems:

Context: Bringing RO treatment directly into homes or buildings to treat municipally supplied tap water or private well water.

Point-of-Use (POU): Typically small under-the-sink RO units providing purified water through a dedicated faucet for drinking and cooking. They usually consist of a sediment pre-filter, an activated carbon pre-filter (for chlorine removal), the RO membrane element, a small storage tank (since RO production is slow), and often a carbon post-filter for taste polishing.

Point-of-Entry (POE): Larger systems installed where the water line enters the house/building, treating all the water used. Less common than POU for RO due to higher cost, maintenance, and water usage (concentrate discharge), but may be used if the entire supply has significant issues (e.g., very high TDS, specific contaminants).

Performance: Effectively remove dissolved salts, heavy metals (lead, arsenic), nitrates, radium, and some organic contaminants, significantly improving taste and purity. Recovery rates have improved in modern units but still generate concentrate water that goes to drain. Require periodic filter and membrane replacement by the homeowner or service provider.

14.2.4 Remineralization: Making RO Water Drinkable

The Issue: RO removes almost all dissolved minerals, including beneficial ones like calcium and magnesium. This highly purified water can be:

- **Corrosive:** Lacking buffering capacity (alkalinity), it can leach metals from pipes and plumbing fixtures.

- **Taste Flat:** Many people find the taste of demineralized water unpalatable.
- **Nutritionally Deficient:** While drinking water isn't the primary source of minerals, complete removal is sometimes debated from a health perspective.

Post-Treatment Step: For municipal and sometimes POU/POE applications, a remineralization (or stabilization) step is added after the RO process. This involves adding back controlled amounts of calcium and magnesium minerals and alkalinity.

Methods:

- Calcite Contactors: Passing the RO permeate through beds of crushed limestone (calcium carbonate) dissolves calcium and adds alkalinity.
- Dolomite Contactors: Similar, using magnesium-containing minerals.
- Chemical Dosing: Injecting solutions like calcium chloride, calcium hydroxide (lime), sodium bicarbonate, and magnesium sulfate. Requires careful dosing control.
- Blending: Mixing a portion of bypassed raw or pretreated water (if suitable quality) with the RO permeate. Less precise control over final composition.

Goal: To produce water that is non-corrosive, has acceptable taste, and meets drinking water quality standards for pH and mineral content.

14.3 Beyond Industrial UPW: Ultrapure Water for Laboratories and Research

While semiconductor fabs are major users of UPW, laboratories across various disciplines (chemistry, biology, medicine, environmental science) also require water of exceptionally high purity for experiments, analyses, and preparing reagents. Contaminants can interfere with sensitive analytical instruments (HPLC, ICP-MS, GC), cell cultures, or chemical reactions.

14.3.1 Diverse Laboratory Water Grades (Type I, II, III):

- Standards organizations (like ASTM, ISO, CLSI) define different grades or types of laboratory water based on parameters like resistivity, TOC, microbial count, and particle levels.
- **Type III Water (RO Grade):** Often produced by single-pass RO, suitable for general lab purposes like glassware rinsing, feeding autoclaves, or as feed for higher purity systems. Resistivity typically > 0.2 Megohm-cm.
- **Type II Water (General Lab Grade):** Higher purity, often produced by RO followed by ion exchange (IX) or EDI, and possibly UV treatment. Suitable for preparing buffers, reagents, and feeding analytical instruments. Resistivity typically > 10-15 Megohm-cm, TOC < 50 ppb.
- **Type I Water (Ultrapure/Reagent Grade):** Highest purity, required for critical applications like HPLC mobile phase preparation, molecular biology (PCR, sequencing), trace metal analysis (ICP-MS), and cell culture. Resistivity > 18.0 Megohm-cm (often 18.2), TOC < 5-10 ppb, very low bacteria and particles, sometimes specific requirements for endotoxins or nucleases.

14.3.2 Benchtop RO/Polishing Systems:

Technology: Laboratories typically use compact, integrated water purification systems designed to produce specific water grades on demand. These systems often incorporate multiple technologies within one unit.

RO's Role: RO serves as the essential primary purification step, removing the bulk of salts, organics, and particles from the tap water feed. This protects and extends the life of the downstream polishing cartridges.

Typical System Configuration (Type I): Tap Water -> Pre-filters (Sediment, Carbon) -> Reverse Osmosis Module -> (Optional Storage Tank) -> Ion Exchange Cartridge(s) -> UV Lamp (for TOC reduction/disinfection) -> Polishing Mixed-Bed IX Cartridge -> Ultrafilter (0.2 or 0.1 μm) at the point of dispense.

Features: These systems provide convenient, on-demand access to high-purity water directly at the lab bench, with integrated monitoring (resistivity, TOC) and easy cartridge replacement.

14.3.3 Specific Research Needs:

Certain research applications may have even more specialized requirements, such as removal of specific endocrine disruptors, RNases/DNases, or extremely low levels of specific ions, requiring tailored polishing steps after the main RO/IX purification.

14.4 Feeding the World: Water for Irrigation

Agriculture is by far the largest consumer of freshwater globally. In many arid and semi-arid regions, traditional water sources are insufficient or becoming saline, threatening food security. Using alternative sources like brackish water or treated wastewater for irrigation is crucial, but salinity poses a major challenge.

14.4.1 The Challenge of Saline Water Sources:

High concentrations of dissolved salts (TDS) in irrigation water can damage crops through several mechanisms:

- **Osmotic Stress:** Makes it harder for plant roots to absorb water.
- **Specific Ion Toxicity:** High levels of certain ions like chloride (Cl^-), sodium (Na^+), and boron (B) are directly toxic to many plants.
- **Soil Degradation:** Accumulation of sodium can damage soil structure (sodicity), reducing water infiltration and aeration.

14.4.2 RO for Removing Salts Harmful to Crops:

Application: RO can desalinate brackish water or treated wastewater to produce water suitable for irrigating sensitive or high-value crops (e.g., fruits, vegetables, greenhouse crops). It effectively removes bulk salts (TDS), reducing osmotic stress, and lowers concentrations of toxic ions like sodium and chloride.

Scale: Can range from relatively small systems for individual farms or greenhouses to larger, centralized plants supplying irrigation districts.

14.4.3 Boron Removal: A Key Challenge for SWRO Irrigation:

The Problem: Boron is an essential micronutrient for plants but becomes toxic at relatively low concentrations (often > 0.5 - 1.0 mg/L for sensitive crops). Seawater contains significant boron levels (around 4-5 mg/L). Standard single-pass SWRO membranes reject boron relatively poorly, especially at neutral pH (boron exists primarily as uncharged boric acid $B(OH)_3$ below pH ~8.5). The permeate may still contain boron levels too high for sensitive crops.

Solutions:

- Second Pass RO at High pH: Increasing the pH of the first-pass permeate (e.g., to pH 9-10) converts boric acid to the charged borate ion ($B(OH)_4^-$), which is much better rejected by the second-pass RO membranes.
- Specialized Boron-Selective Membranes: Membranes engineered for enhanced boron rejection are available, potentially reducing the need for extensive pH adjustment or a second pass in some cases.
- Ion Exchange: Specialized boron-selective ion exchange resins can be used as a polishing step after RO.

Significance: Boron removal adds complexity and cost to using SWRO permeate for irrigation, limiting its widespread application primarily to high-value agriculture or areas with extreme water scarcity.

14.4.4 Economic and Sustainability Considerations:

- **Cost:** Desalinating water for irrigation using RO is generally much more expensive than using traditional freshwater sources. Its economic viability often depends on the value of the crop being grown, government subsidies, and the lack of alternative water sources.

- **Energy:** Energy consumption remains a significant factor, although advancements continue to lower it. Use of renewable energy sources (solar) is being explored for agricultural RO systems.
- **Concentrate Management:** Disposing of the saline concentrate inland can be a major environmental and logistical challenge.
- **Water Quality Matching:** It's sometimes possible to blend RO water with lower quality water to achieve an acceptable salinity level for specific crop tolerances, optimizing water use and reducing costs.

14.5 Pharmaceutical Water: Beyond PW and WFI Generation

While Chapter 13 discussed the core production of USP Purified Water (PW) and Water For Injection (WFI) using RO, the need for high-purity water permeates many other aspects of pharmaceutical operations.

14.5.1 Water Quality for Specific Processes (e.g., Diagnostics):

Manufacturing diagnostic reagents, calibration standards, or specific assays may require water meeting unique purity specifications beyond standard PW, perhaps with extremely low levels of specific interfering ions, organics, or particles. RO often forms the basis of the purification train for these custom water needs.

14.5.2 Final Rinsing and Cleaning Validation:

Water used for the final rinse of manufacturing equipment (tanks, pipes, filling lines) must be of very high purity (often PW or WFI quality) to ensure no residues are left behind. RO is critical in providing this rinse water.

Cleaning validation procedures often involve analyzing rinse water samples (using TOC and conductivity) to demonstrate that equipment is clean; the purity of the initial rinse water itself must be impeccable.

14.5.3 Laboratory Use in Pharma QC/R&D:

Similar to general labs (Section 14.3), pharmaceutical Quality Control (QC) and Research & Development (R&D) labs rely heavily on Type I and Type II water produced by systems utilizing RO as the primary purification step. This ensures accuracy and reliability in drug analysis, formulation development, and research experiments.

14.6 Other Niche Applications

RO's versatility extends to many other specialized areas, including:

- **Kidney Dialysis:** Water used to prepare dialysis fluid (dialysate) must meet stringent purity standards (AAMI standards) to prevent contaminants from entering the patient's bloodstream. Central RO systems in hospitals or clinics provide this purified water, typically followed by UV and ultrafiltration. Portable RO units are used for home dialysis.
- **Car Washes:** Spot-free rinsing requires water with very low TDS to prevent mineral spotting as water evaporates. Small RO systems are commonly used to produce this final rinse water.
- **Humidification:** Using RO water in humidifiers (for industrial processes, cleanrooms, or comfort) prevents scaling and the dispersal of mineral dust.
- **Aquariums and Aquaculture:** Maintaining specific water chemistry is vital for sensitive aquatic life. RO/DI (Deionization) systems are used by hobbyists and commercial aquariums to remove harmful contaminants from tap water before adding back specific salts and buffers. Aquaculture facilities may use RO to treat source water or recycle process water.
- **Maple Sap Concentration:** Similar to juice concentration, RO can remove water from maple sap before boiling it down

into syrup, significantly saving time and energy during the evaporation process.

- **Hydrogen Production (Electrolysis Feedwater):** Water electrolysis to produce green hydrogen requires very high purity feedwater to prevent electrode fouling and ensure efficiency. RO/EDI is commonly used.

14.7 Chapter Summary and Key Takeaways

- RO technology is adapted for numerous specialized applications beyond large-scale industry.
- Potable Water Production is a major application, from large municipal SWRO/BWRO plants ensuring water security, to small community systems, down to POU units improving home water quality. Remineralization is often needed for RO-treated drinking water.
- Laboratory Water Purification relies on RO as the primary step in integrated systems producing various grades (Type I, II, III) of ultrapure water essential for research and analysis.
- Irrigation using RO-treated saline water is technically feasible but faces economic challenges and specific technical hurdles like boron removal from seawater.
- Pharmaceutical applications extend beyond bulk PW/WFI to include specialized process water, critical cleaning rinse water, and laboratory QC/R&D needs.
- Other niche applications include kidney dialysis, car washes, humidification, aquariums, maple sap concentration, and hydrogen production, showcasing RO's adaptability.
- Each specialized application has unique requirements regarding scale, purity, post-treatment, and operational considerations, demonstrating the tailored nature of RO system deployment.

14.8 References

1. World Health Organization (WHO). Guidelines for Drinking-water Quality. (Provides global standards for potable water).

2. AWWA. Various Standards and Manuals related to Drinking Water Treatment and Desalination.
3. National Sanitation Foundation (NSF) International. Standards for Drinking Water Treatment Units (e.g., NSF/ANSI 58 for RO Systems).
4. ASTM International. Standard Specification for Reagent Water (D1193). (Defines laboratory water types).
5. Clinical and Laboratory Standards Institute (CLSI). CLSI GP40 - Preparation and Testing of Reagent Water in the Clinical Laboratory.
6. Shannon, M. A., Bohn, P. W., Elimelech, M., Georgiadis, J. G., Marĩas, B. J., & Mayes, A. M. (2008). Science and technology for water purification in the coming decades. Nature, 452(7185), 301-310. (Discusses future water needs, including agriculture).
7. Yermiyahu, U., Ben-Gal, A., Keren, R., & Reid, R. J. (2007). Combined effect of salinity and excess boron on plant growth and yield. Plant and Soil, 290(1-2), 191-202. (Example of boron toxicity research).
8. Association for the Advancement of Medical Instrumentation (AAMI). ANSI/AAMI RD52: Dialysate for hemodialysis. (Standards for dialysis water).
9. Perkins, T. D. (2002). Using reverse osmosis in the sugarhouse. Maple Syrup Journal, 2(1), 12-15. (Example of niche application).

Chapter 15: Water Quality Parameters and Standards - Understanding the Target and the Medium

15.1 Introduction: The Importance of Knowing Your Water

Water is often called the "universal solvent," capable of dissolving or suspending a vast array of substances. Its composition varies dramatically depending on its source (rain, river, lake, groundwater, seawater, wastewater) and its journey through the environment and human systems. From the perspective of Reverse Osmosis, water is both the medium we work with and the source of challenges we must overcome.

A thorough understanding of water quality is paramount for anyone involved with RO systems. It dictates:

- The feasibility of using RO for a specific application.
- The design of the necessary pretreatment train.
- The prediction of RO system performance (flux, rejection, pressure).
- The potential for fouling and scaling.
- The selection of appropriate materials of construction.
- The need for post-treatment steps.
- Compliance with regulatory standards for the final product water.

This chapter serves as a reference guide to the most important water quality parameters encountered in RO applications. We'll define each parameter, explain its significance in the context of RO operation and membrane performance, discuss common measurement units, and briefly touch upon typical analytical methods. We will also place these parameters in the broader context of drinking water standards to understand the quality targets often aimed for in potable water applications.

15.2 Why Water Quality Matters for RO Systems

As we've seen throughout this book, RO membranes are sensitive components. Their performance and longevity are directly impacted by the chemistry and physical characteristics of the feed water:

- **Scaling & Fouling:** Many parameters directly contribute to scaling (Hardness, Alkalinity, Silica, Sulfate, etc.) or fouling (Suspended Solids, Organics, Microbes). Knowing their concentrations is essential for designing preventative measures (pretreatment, chemical dosing).
- **Osmotic Pressure & Energy:** TDS dictates the osmotic pressure, which determines the minimum feed pressure and thus the energy required for separation.
- **Membrane Performance:** pH affects the charge on the membrane surface and the ionization state of certain solutes (like silica), influencing rejection. Temperature significantly impacts water flux.
- **Material Compatibility:** High chloride levels require corrosion-resistant materials. Oxidants (measured indirectly via ORP) dictate the need for dechlorination to protect TFC membranes.
- **Product Quality:** The concentration of specific ions or contaminants in the feed water, combined with the membrane's rejection capability, determines if the final permeate will meet the quality standards for its intended use (drinking water, boiler feed, UPW, etc.).

15.3 Key Water Quality Parameters and Their Significance for RO

Let's examine the parameters most frequently encountered and their relevance:

15.3.1 pH: The Master Variable

Definition: A measure of the acidity or basicity of water on a logarithmic scale from 0 (highly acidic) to 14 (highly basic), with 7 being neutral. It reflects the hydrogen ion (H^+) concentration.

Significance for RO:

- **$CaCO_3$ Scaling:** Solubility of calcium carbonate decreases dramatically as pH increases above ~7.5. LSI calculations heavily depend on pH. Acid dosing lowers pH to prevent $CaCO_3$ scale.
- **Silica Scaling:** Solubility of silica increases significantly above pH ~8.5. Sometimes pH is deliberately raised (e.g., between RO passes) to improve silica rejection and prevent scaling.
- **Membrane Stability:** Membranes have specific operating pH limits (e.g., pH 2-11 for TFC, narrower for CA). Cleaning often involves operating near these limits (e.g., pH 2-3 or pH 11-12).
- **Rejection of Weak Acids/Bases:** The rejection of species like silica, boron, ammonia, and some organics depends on their ionization state, which is controlled by pH.
- **Corrosion:** Low pH water is more corrosive.

Measurement: pH meter with electrode (requires regular calibration).

15.3.2 Total Dissolved Solids (TDS): The Overall Salt Load

Definition: The total mass concentration of all dissolved inorganic and organic substances in water.

Units: Milligrams per liter (mg/L) or Parts per million (ppm) (practically equivalent for dilute solutions).

Significance for RO:

- **Osmotic Pressure:** Directly determines the osmotic pressure that must be overcome. Higher TDS requires higher feed pressure.
- **Permeate Quality:** Indicates the overall contaminant load the RO must remove. The expected permeate TDS is roughly Feed TDS * (1 - % Rejection).
- **General Classification:** Used to categorize water (e.g., Freshwater < 1,000 ppm; Brackish 1,000-15,000 ppm; Saline/Seawater > 15,000-30,000 ppm).

Measurement: Gravimetrically (evaporating a filtered sample and weighing the residue) or estimated from conductivity.

15.3.3 Conductivity: An Electrical Proxy for TDS

Definition: A measure of water's ability to conduct electricity, which is proportional to the concentration of dissolved ions.

Units: Microsiemens per centimeter (µS/cm) or Millisiemens per centimeter (mS/cm).

Significance for RO:

- **Quick TDS Estimate:** Provides a rapid, inexpensive estimate of TDS. The conversion factor (TDS ≈ Conductivity * Factor) varies with water composition (typically 0.5-0.8), but conductivity is excellent for tracking relative changes.
- **Performance Monitoring:** Permeate conductivity is the primary real-time indicator of RO rejection performance. An increase signals a problem.

Measurement: Conductivity meter/probe (requires temperature compensation).

15.3.4 Suspended Solids (TSS & Turbidity): The Physical Cloggers

Definition: Undissolved particulate matter (TSS) and the cloudiness it causes (Turbidity). Colloids are the very fine fraction.

Units: TSS in mg/L; Turbidity in NTU (Nephelometric Turbidity Units). SDI (unitless index) measures fouling potential.

Significance for RO:

Major cause of physical fouling (clogging feed channels, coating surfaces). Requires effective pretreatment (filtration, clarification, MF/UF) to achieve low turbidity (<0.1-0.2 NTU) and low SDI (<3-5).

Measurement: TSS by filtration/weighing; Turbidity by nephelometer; SDI by standard filter plugging test.

15.3.5 Hardness: The Primary Scaling Ions (Ca^{2+}, Mg^{2+})

Definition: Primarily the concentration of dissolved calcium (Ca^{2+}) and magnesium (Mg^{2+}) ions.

Units: Usually expressed as equivalent calcium carbonate (mg/L as $CaCO_3$).

Significance for RO:

Calcium is a key component of the most common scales ($CaCO_3$, $CaSO_4$). Magnesium can contribute to scaling ($Mg(OH)_2$, $MgSiO_3$) under certain conditions. Hardness levels are critical inputs for scaling prediction (LSI, etc.) and determining antiscalant needs or the potential benefit of upstream softening.

Measurement: Titration (e.g., with EDTA) or individual ion analysis (ICP, AA).

15.3.6 Alkalinity: Buffering Capacity and Carbonate Source

Definition: Water's capacity to neutralize acids. Primarily due to bicarbonate (HCO_3^-), carbonate (CO_3^{2-}), and hydroxide (OH^-) ions.

Units: Usually expressed as equivalent calcium carbonate (mg/L as $CaCO_3$).

Significance for RO:

- **$CaCO_3$ Scaling:** Bicarbonate/carbonate ions are the other key ingredient (along with calcium and high pH) for calcium carbonate scale. Alkalinity is a critical input for LSI calculation.
- **pH Buffering:** High alkalinity water resists changes in pH. This affects acid dosing requirements if pH adjustment is used.
- **CO_2 in Permeate:** Alkalinity relates to dissolved CO_2. Acid addition converts alkalinity to CO_2, which passes through the RO membrane.

Measurement: Titration with standard acid to specific pH endpoints (e.g., pH 8.3 and pH 4.5).

15.3.7 Specific Ions: Targeted Concerns

While TDS gives the total picture, the concentration of specific ions is often crucial:

15.3.7.1 Chloride (Cl^-) and Sulfate (SO_4^{2-}):

Major anions contributing to TDS and osmotic pressure. High chloride levels (> few hundred ppm) increase corrosivity towards stainless steels (requiring Duplex/Super Duplex). Sulfate is a component of calcium sulfate and barium/strontium sulfate scales.

15.3.7.2 Sodium (Na^+) and Potassium (K^+):

Major cations contributing to TDS and osmotic pressure. Generally well rejected and don't typically cause scaling issues themselves (except in very specific ZLD contexts).

15.3.7.3 Nitrate (NO_3^-) and Fluoride (F^-):

Health-related contaminants in drinking water often targeted for removal by RO, especially in groundwater. Maximum Contaminant Levels (MCLs) exist in drinking water regulations.

15.3.7.4 Iron (Fe) and Manganese (Mn):

Can exist in dissolved forms (Fe^{2+}, Mn^{2+}) in anoxic groundwater. If oxidized (by air exposure or disinfectants), they precipitate as insoluble hydroxides/oxides ($Fe(OH)_3$, MnO_2) causing severe fouling (reddish/black deposits). Require specific pretreatment removal (oxidation + filtration, specialized media). Even low levels (<0.05 ppm) can be problematic.

15.3.7.5 Heavy Metals (e.g., Arsenic, Lead, Mercury, Cadmium, Chromium):

Toxic contaminants regulated in drinking water. RO is generally effective at removing dissolved heavy metals, often used specifically for arsenic removal from groundwater.

Measurement (Specific Ions): Ion Chromatography (IC), Inductively Coupled Plasma (ICP), Atomic Absorption (AA), ion-selective electrodes.

15.3.8 Silica (SiO_2): The Stealthy Scalant

Definition: Dissolved silica, exists as silicic acid (H_4SiO_4 or $Si(OH)_4$) at neutral pH.

Significance for RO:

Solubility is limited (typically ~120-150 mg/L at 25°C, depending on pH and other ions) and polymerization/precipitation can occur in the concentrate stream, forming tenacious scale. Rejection by RO is pH-dependent (better at high pH). Often limits recovery in treating certain groundwaters or industrial effluents. Requires careful monitoring and management via recovery limits, pH adjustment, or specialized antiscalants.

Measurement: Colorimetric methods (molybdate blue/yellow), ICP.

15.3.9 Organic Matter (TOC, Color, etc.): Fouling Potential

Definition: Natural Organic Matter (NOM), synthetic organics, oils/greases. Measured by TOC (Total Organic Carbon), Color, UVA_{254}, etc.

Significance for RO:

Major cause of organic fouling (adsorption) and can enhance biofouling (provides nutrients). Requires pretreatment (coagulation, carbon adsorption, UF) especially for surface waters.

Measurement: TOC analyzer, spectrophotometer (Color, UV), specific GC/MS analysis for synthetic organics if needed.

15.3.10 Microorganisms: Biofouling and Health Risks

Definition: Bacteria, algae, fungi, viruses, protozoa. Measured indirectly via plate counts (HPC), ATP, or specific tests.

Significance for RO:

Cause biofouling (slime formation), leading to high ΔP and flux loss. If producing drinking water or pharmaceutical water, removal/inactivation is critical for health. Requires disinfection strategies and biofilm control.

Measurement: Microbiological plating techniques, ATP luminescence, microscopic analysis.

15.3.11 Temperature: The Kinetic Modifier

Definition: Water temperature.

Units: Degrees Celsius (°C) or Fahrenheit (°F).

Significance for RO:

- **Viscosity & Flux:** Higher temperature decreases viscosity, significantly increasing water flux.
- **Salt Passage:** Higher temperature slightly increases salt diffusion, potentially decreasing rejection.
- **Scaling:** Affects solubility of scalants (e.g., $CaCO_3$ less soluble, $CaSO_4/SiO_2$ more soluble at higher temps).
- **Chemical Reactions:** Affects reaction rates (disinfection, cleaning).
- **Essential for Normalization:** Temperature MUST be known to normalize performance data accurately.

Measurement: Thermometer, RTD, thermocouple.

15.3.12 Oxidation-Reduction Potential (ORP): Assessing Chemical State

Definition: A measure of water's tendency to oxidize or reduce substances. Reflects the balance between oxidizing agents (like chlorine, oxygen) and reducing agents (like H_2S, SBS).

Units: Millivolts (mV).

Significance for RO:

Primarily used to ensure complete removal of chlorine or other oxidants before TFC membranes. High positive ORP (> +200 mV) indicates presence of oxidants harmful to polyamide membranes. After SBS dosing, ORP should drop significantly (often below +150 mV). Negative ORP can indicate reducing conditions (e.g., presence of H_2S).

Measurement: ORP meter/probe (similar to pH probe).

15.4 Measuring the Parameters: Sampling and Analysis Considerations

- **Representative Sampling:** Ensure samples accurately reflect the water at the specific point in the system being assessed (e.g., raw water, RO feed after pretreatment, permeate, concentrate). Use appropriate clean sample containers.

- **Timeliness:** Some parameters change rapidly after sampling (pH, dissolved gases, ORP, chlorine residual). Measure these on-site immediately if possible.
- **Lab Quality:** Use accredited laboratories experienced in water analysis for critical parameters.
- **Frequency:** Monitoring frequency depends on source water variability and system criticality (e.g., continuous online monitoring for permeate conductivity, daily SDI, periodic full analysis).

15.5 Putting it in Context: Drinking Water Quality Standards

While RO is used for many applications, producing potable water is a primary one. Drinking water standards provide benchmarks for safety and aesthetic quality.

15.5.1 The Role of Standards (Health, Aesthetics, Operations):

- **Health Protection:** Set legally enforceable limits (Maximum Contaminant Levels - MCLs) for contaminants known to pose health risks (e.g., pathogens, toxic chemicals like arsenic, nitrates).
- **Aesthetic Quality:** Recommend limits (Secondary MCLs or guidelines) for parameters affecting taste, odor, color, or appearance (e.g., TDS, chloride, sulfate, iron, manganese).
- **Operational Guidance:** Some standards relate to corrosivity or scaling potential, relevant for distribution systems.

15.5.2 World Health Organization (WHO) Guidelines:

Provide internationally recognized, health-based guideline values for numerous chemical and microbial contaminants. While not legally binding in themselves, they form the basis for many national standards.

Emphasize a Water Safety Plan approach (risk assessment and management from catchment to consumer).

15.5.3 Examples of National Standards (e.g., US EPA):

- Countries set their own legally enforceable standards. In the USA, the Environmental Protection Agency (EPA) sets Primary (health-related, mandatory) and Secondary (aesthetic, non-mandatory) MCLs under the Safe Drinking Water Act.
- Examples (US EPA Primary MCLs relevant to RO): Arsenic (10 μg/L), Nitrate (10 mg/L as N), Fluoride (4.0 mg/L), Lead (Action Level 15 μg/L), certain Radionuclides.
- Examples (US EPA Secondary MCLs): Chloride (250 mg/L), Sulfate (250 mg/L), TDS (500 mg/L), Iron (0.3 mg/L), Manganese (0.05 mg/L), pH (6.5-8.5).

15.5.4 How RO Performance Relates to Standards:

- RO is highly effective at removing most contaminants with MCLs (TDS, salts, heavy metals, nitrates, radionuclides).
- RO permeate often has very low TDS, potentially requiring remineralization (Section 14.2.4) to meet secondary standards for taste and corrosivity (pH, alkalinity).
- RO provides an excellent barrier against microbial contaminants, but maintaining distribution system integrity and sometimes final disinfection are still needed.
- Designers use feed water quality and target standards to select appropriate membranes and predict if permeate quality will meet requirements, potentially necessitating multiple passes or post-treatment.

15.6 Chapter Summary and Key Takeaways

- Understanding water quality parameters is fundamental to designing, operating, and troubleshooting RO systems.
- Key parameters influence scaling/fouling potential (Hardness, Alkalinity, Silica, SO_4^{2-}, Fe, Mn, Organics, Microbes, TSS/Turbidity/SDI), osmotic pressure/energy

(TDS/Conductivity), membrane stability/performance (pH, Temperature, ORP), and material selection (Cl^-, pH).

- Accurate sampling and analysis are crucial; rely on calibrated instruments and reputable labs.
- Drinking Water Standards (WHO, national like EPA) set health-based (MCLs) and aesthetic limits that often drive the need for RO or define the target quality for RO permeate.
- RO is effective at meeting most MCLs but often requires post-treatment (remineralization) to meet secondary standards and operational needs related to taste and corrosivity.
- A comprehensive water analysis is the essential starting point for any RO project.

15.7 References

1. American Public Health Association (APHA), American Water Works Association (AWWA), & Water Environment Federation (WEF). Standard Methods for the Examination of Water and Wastewater (Multiple Editions). (The definitive reference for analytical methods).
2. AWWA. Water Quality and Treatment: A Handbook on Drinking Water (6th ed.). (Comprehensive text on water quality parameters and treatment).
3. World Health Organization (WHO). Guidelines for Drinking-water Quality (4th ed. incorporating first addendum, 2017). (Global reference for health-based guidelines).
4. United States Environmental Protection Agency (US EPA). National Primary Drinking Water Regulations & National Secondary Drinking Water Regulations. (Specific US standards).
5. Snoeyink, V. L., & Jenkins, D. (1980). Water Chemistry. John Wiley & Sons. (Classic textbook on aquatic chemistry principles).
6. Faust, S. D., & Aly, O. M. (1998). Chemistry of Water Treatment (2nd ed.). CRC Press. (Covers chemical principles relevant to treatment processes).

7. Manufacturer Technical Literature (Dow, Hydranautics, Toray, etc.): Often include tables of typical feed water parameters and their significance for RO.

Chapter 16: Chemical Considerations in RO Systems - Scaling, Corrosion, and Compatibility

16.1 Introduction: Chemistry Under Pressure

We've meticulously designed the RO system, selected membranes, configured arrays, and implemented a pretreatment train to deliver water meeting specific quality guidelines. However, the chemical interactions don't stop once the water enters the high-pressure pump. The very nature of Reverse Osmosis – removing pure water and concentrating the remaining dissolved solids – creates a unique and dynamic chemical environment within the membrane elements and downstream piping.

As the concentration factor (Beta) increases along the flow path, salts that were stable in the feed water can become supersaturated, leading to precipitation or scaling directly on the membrane surfaces or feed spacers. Simultaneously, the combination of high pressure, specific ion concentrations (especially chlorides), temperature, and flow can create conditions conducive to corrosion of metallic components like pumps, piping, and pressure vessels. Furthermore, all materials used in the system – membranes, pipes, valves, seals, tanks – must be chemically compatible not only with the feed water but also with the concentrated brine, cleaning chemicals, and any dosed treatment chemicals (like antiscalants or bisulfite).

Ignoring these internal chemical considerations can lead to premature membrane failure, unexpected leaks, equipment degradation, and costly downtime. This chapter delves deeper into the chemistry behind the most common scaling phenomena, explores the mechanisms and mitigation of corrosion within RO systems, and highlights the critical importance of ensuring chemical compatibility across all system components.

16.2 The Inevitable Challenge: Scaling Chemistry Revisited

We've discussed scaling as a major type of fouling and the use of indices like LSI to predict it. Here, we look more closely at the chemistry driving the formation of the most common scales within the RO membrane environment.

16.2.1 Supersaturation: The Driving Force

Scaling occurs when the concentration of dissolved ions forming a specific mineral salt exceeds the solubility limit of that salt under the prevailing conditions (temperature, pH, ionic strength) within the RO system, particularly in the boundary layer near the membrane surface and in the bulk concentrate stream. This state is called supersaturation. The higher the degree of supersaturation, the greater the driving force for precipitation (scaling). RO inherently creates supersaturation by removing water and increasing the concentration factor (β).

16.2.2 Calcium Carbonate ($CaCO_3$) Scaling: The Usual Suspect

16.2.2.1 Equilibrium and pH Dependence:

This is the most frequently encountered scale in non-seawater RO. Its formation is governed by a complex equilibrium involving calcium ions (Ca^{2+}), bicarbonate ions (HCO_3^-), carbonate ions (CO_3^{2-}), dissolved carbon dioxide (CO_2), and water. A simplified representation often considered is:

$$Ca^{2+} + 2HCO_3^- \rightleftharpoons CaCO_3(s) + CO_2(aq) + H_2O$$

The key takeaway is that this equilibrium is highly sensitive to pH. As pH increases:

- Bicarbonate shifts towards carbonate ($HCO_3^- \rightarrow CO_3^{2-} + H^+$).
- The concentration of carbonate ions (CO_3^{2-}), a direct precursor to $CaCO_3$ scale, increases significantly.

- The equilibrium shifts to the right, favoring the precipitation of solid $CaCO_3$.

Conversely, lowering the pH (adding acid) shifts the equilibrium to the left, dissolving $CaCO_3$ or preventing its formation. Temperature also plays a role; $CaCO_3$ becomes less soluble as temperature increases (unlike many salts).

16.2.2.2 LSI and S&DSI as Predictive Tools:

Because of the complex equilibrium, simple solubility calculations are difficult. The Langelier Saturation Index (LSI) and the Stiff & Davis Stability Index (S&DSI) (which adjusts for higher ionic strength) are widely used empirical indices calculated from Ca^{2+}, Alkalinity, TDS, pH, and Temperature (as discussed in Ch 3 & 7). They provide a practical measure of the driving force for $CaCO_3$ precipitation in the concentrate stream. A positive index indicates a tendency to scale.

16.2.3 Calcium Sulfate ($CaSO_4$) Scaling: The Hard Scale

16.2.3.1 Solubility Limits and Ion Products:

Calcium sulfate scale (commonly gypsum, $CaSO_4 \cdot 2H_2O$) formation is primarily governed by exceeding its solubility product constant (Ksp). The Ion Product (IP) is calculated from the molar concentrations of calcium and sulfate ions in the concentrate: $IP = [Ca^{2+}] * [SO_4^{2-}]$. If IP exceeds Ksp under the specific temperature and ionic strength conditions, precipitation is likely. Unlike $CaCO_3$, $CaSO_4$ solubility is much less dependent on pH within typical RO operating ranges.

16.2.3.2 Temperature Effects:

Calcium sulfate exhibits retrograde solubility at lower temperatures relevant to RO, meaning it becomes more soluble as temperature increases (opposite to $CaCO_3$). This means scaling risk might be slightly higher in colder feed water conditions for the same ionic concentrations.

Prediction: Design software calculates the % saturation (IP / Ksp * 100%) in the concentrate. Typically, designs aim to keep $CaSO_4$

saturation below 100% if no specific antiscalant is used, or up to limits specified by the antiscalant vendor (e.g., 200-400% or higher, depending on the product).

16.2.4 Silica (SiO_2) Scaling: The Polymerization Problem

16.2.4.1 Mechanisms and Influencing Factors:

Silica scaling is complex and doesn't follow simple precipitation rules. Dissolved silica exists mainly as monomeric silicic acid ($Si(OH)_4$) at neutral pH. As concentration increases, it begins to polymerize, forming chains and eventually three-dimensional colloidal silica gels that deposit on the membrane. This process is influenced by:

- **Concentration:** Polymerization rate increases sharply above silica's solubility limit (around 120-150 mg/L at 25°C, but can vary).
- **pH:** Polymerization is slowest around pH 7-8. Below pH 7, it increases slightly. Above pH ~8.5-9, silica solubility increases significantly as silicic acid deprotonates to form silicate ions ($HSiO_3^-$), reducing the scaling tendency.
- **Temperature:** Higher temperatures generally increase silica solubility and can slow polymerization kinetics.
- **Presence of Metal Hydroxides:** Iron and aluminum hydroxides can co-precipitate with silica or act as nucleation sites.
- **Hardness Ions (Ca^{2+}, Mg^{2+}):** Can sometimes form metal silicates, especially at higher pH.

16.2.4.2 Challenges in Prediction and Removal:

Predicting silica scaling accurately is difficult due to the complex polymerization kinetics. Design software uses empirical limits based on concentration and temperature (often limiting concentrate SiO_2 to < 150-180 mg/L, although specialized antiscalants may allow higher limits). Silica scale, once formed, is often glassy and very difficult to remove with standard acid or alkaline cleaners. High pH (~12) and high temperature (~40°C) cleaning, sometimes with specific additives like ammonium bifluoride (use with extreme caution and check membrane compatibility!), may be required. Prevention through conservative recovery limits is often the best strategy.

6.2.5 Other Potential Scales ($BaSO_4$, $SrSO_4$, $Ca_3(PO_4)_2$)

Barium Sulfate ($BaSO_4$) & Strontium Sulfate ($SrSO_4$): These are extremely insoluble. Even if Barium (Ba^{2+}) and Strontium (Sr^{2+}) are present only in trace amounts (ppb or low ppm) in the feed, they can easily exceed their solubility limits at typical RO concentration factors, forming very hard, tenacious scale that is nearly impossible to remove chemically. Accurate feed analysis for Ba and Sr is critical if high recovery is planned, and specific antiscalants targeting these scales may be needed. Limits are often kept below 100% saturation even with antiscalants.

Calcium Phosphate ($Ca_3(PO_4)_2$): Can precipitate if sufficient phosphate (PO_4^{3-}) is present in the feed water (e.g., from agricultural runoff, wastewater effluent, or corrosion inhibitor residuals). Solubility is pH-dependent (less soluble at higher pH).

16.2.6 How Antiscalants Interfere: A Chemical Perspective

Antiscalants (Chapter 9) don't remove scaling ions, but rather allow the system to operate under supersaturated conditions without forming scale deposits. They work primarily through:

- **Threshold Inhibition:** Adsorbing onto the surface of microscopic crystal nuclei (the initial seeds of scale formation), preventing them from growing larger. Requires only very low concentrations (ppm level).
- **Crystal Modification:** Interfering with the regular lattice structure of growing crystals, distorting their shape and making them less likely to adhere to surfaces or form dense scale. Distorted crystals may be softer and more easily flushed away.

- **Dispersion:** Adsorbing onto colloidal particles or crystals, increasing their negative charge and causing them to repel each other, preventing agglomeration and deposition.

16.3 The Slow Decay: Corrosion in RO Systems

While scaling involves adding material, corrosion involves removing it – the gradual degradation of metallic components through electrochemical reactions with their environment (the water).

16.3.1 Why Corrosion Matters (Leaks, Contamination, Costs):

- **Equipment Failure:** Can lead to leaks in piping, pressure vessels, pump casings, and valve bodies, causing system downtime and costly repairs/replacements.
- **Product Contamination:** Corrosion products (e.g., iron oxides, copper ions) can contaminate the feed water, potentially fouling membranes or compromising permeate quality.
- **Safety Hazards:** Failure of high-pressure components due to corrosion poses a significant safety risk.

16.3.2 Common Forms of Corrosion in RO Systems:

16.3.2.1 General (Uniform) Corrosion:

Relatively uniform metal loss across the surface. Less common in RO systems using appropriate stainless steels, but could occur with carbon steel components exposed to low pH or oxygenated water.

16.3.2.2 Pitting Corrosion (The Chloride Menace):

Localized attack creating small holes or pits. This is the primary concern for stainless steels (even 316L) in the presence of chloride ions (Cl^-). Chlorides disrupt the passive oxide layer that normally protects stainless steel, leading to rapid, localized penetration. Risk increases significantly with higher chloride concentration, higher temperature, and lower pH.

16.3.2.3 Crevice Corrosion:

Localized corrosion occurring within stagnant areas or crevices (e.g., under gaskets, seals, deposits, or at flange joints) where the local chemistry can become more aggressive (e.g., lower pH, higher chloride concentration) than the bulk water.

16.3.2.4 Galvanic Corrosion:

Occurs when two dissimilar metals are in electrical contact in a conductive electrolyte (water). The more "active" metal corrodes preferentially to protect the more "noble" metal. Requires careful material selection to avoid incompatible metal pairings (e.g., stainless steel coupled directly with carbon steel or brass in saline water).

16.3.3 Key Factors Influencing Corrosion Rates:

16.3.3.1 Materials Selection (Stainless Steels, Alloys, Plastics):

The single most important factor. Using materials resistant to the specific water chemistry is crucial (See Section 16.4).

16.3.3.2 Water Chemistry:

- **Chlorides (Cl^-):** The main driver for pitting/crevice corrosion of stainless steels. Higher levels require more resistant alloys (Duplex, Super Duplex).
- **pH:** Low pH (< 6.5) increases general corrosion rates for most metals. High pH (> 8.5) can passivate some metals but might increase risk for others (e.g., aluminum).
- **Dissolved Oxygen (O_2):** Often required for corrosion reactions (acts as cathodic reactant), but also helps form passive layers on stainless steel. Complete removal isn't usually practical or necessary in RO feed.
- **TDS/Conductivity:** Higher conductivity increases the rate of electrochemical corrosion cells.
- **Other Species:** Sulfides (H_2S), ammonia (NH_3), carbonates can also influence corrosion.

16.3.3.3 Temperature and Flow Velocity:

Higher temperatures generally accelerate corrosion reaction rates. Very high flow velocities can cause erosion-corrosion, while very low/stagnant flow promotes crevice corrosion and allows deposits to form.

16.3.4 Mitigation Strategies:

- **Proper Material Selection:** Choose alloys (316L, Duplex, Super Duplex) appropriate for the chloride level, temperature, and pressure (Chapter 6). Use non-metallic components (FRP vessels, plastic piping for low pressure) where feasible.
- **Water Chemistry Control:** Maintain pH within acceptable ranges. While complete deoxygenation isn't typical for RO feed, avoiding excessive aeration can help. Proper pretreatment to remove corrosive species like H_2S.
- **Design:** Minimize crevices. Avoid dissimilar metal contacts or use insulating gaskets/fittings. Ensure proper flow velocities.
- **Operational Practices:** Regular flushing to prevent stagnation. Prompt leak repairs.

16.4 Choosing Wisely: Chemical Compatibility of RO Materials

Beyond corrosion of metals, all wetted materials in the RO system must resist degradation from the process water and any introduced chemicals (pretreatment, cleaning).

16.4.1 Membrane Compatibility: Beyond Feed Water

16.4.1.1 Oxidants (Chlorine, etc.):

As repeatedly stressed, standard TFC polyamide membranes have extremely poor tolerance to free chlorine (<0.1 ppm continuous). Other oxidants like ozone, peroxide, permanganate are also generally incompatible. This dictates the necessity of dechlorination (SBS,

GAC). Even trace amounts cause irreversible damage to the salt rejection layer. Cellulose Acetate (CA) has limited tolerance (0.5-1.0 ppm free chlorine).

16.4.1.2 pH Limits:

Membranes have specified pH ranges for continuous operation (e.g., pH 2-11 for many TFCs) and often slightly wider ranges for short-term cleaning (e.g., pH 1-13 for some). Operating outside these limits causes hydrolysis (chemical breakdown) of the polymer, leading to irreversible performance loss.

16.4.1.3 Temperature Limits:

Exceeding the maximum operating temperature (e.g., ~45°C for standard TFCs) can cause irreversible compaction or degradation. Cleaning temperatures also have limits.

16.4.1.4 Cleaning Chemical Compatibility:

Only use cleaning chemicals specifically approved by the membrane manufacturer. Some solvents, strong surfactants, or incompatible biocides can damage the membrane polymer or adhesives used in element construction. Always check compatibility lists.

16.4.2 Piping, Vessels, and Valves: Metals vs. Plastics

16.4.2.1 Stainless Steel Grades and Chloride Limits:

The choice between 304SS, 316L SS, Duplex SS, and Super Duplex SS for high-pressure piping and pump components depends almost entirely on the maximum chloride concentration and temperature expected in the feed/concentrate streams. Using an inadequate grade leads inevitably to pitting/crevice corrosion.

16.4.2.2 PVC, CPVC, PP, PVDF Limitations:

Used for low-pressure piping (permeate, drains, chemical lines). Have excellent resistance to many aqueous solutions but have temperature, pressure, and UV limitations. They are generally incompatible with certain organic solvents. Check chemical resistance charts for specific compatibility with cleaning chemicals or non-aqueous streams.

16.4.3 Elastomers (Seals and Gaskets): The Weak Link?

O-rings used in element interconnectors, end cap seals, and gaskets in valve seats are critical for preventing leaks. Their material must be compatible with the water and chemicals.

16.4.3.1 Common Materials:

- **EPDM (Ethylene Propylene Diene Monomer):** Good resistance to water, steam, dilute acids/alkalis, polar solvents. Standard for general water service O-rings. Poor resistance to oils and hydrocarbon solvents.
- **Buna-N (Nitrile):** Good resistance to oils, petroleum products. Moderate resistance to water. Less common as primary RO seal material than EPDM.
- **Viton® (FKM - Fluoroelastomer):** Excellent resistance to high temperatures, oils, fuels, many acids, and some solvents. Often used in more demanding chemical environments or where oils might be present. More expensive.

16.4.3.2 Compatibility:

Check compatibility with feed water components (e.g., oils), cleaning chemicals (some detergents or solvents can attack EPDM), and any disinfection/sanitization agents (e.g., ozone can degrade EPDM and Buna-N; Viton® is more resistant). Incorrect elastomer choice is a common cause of leaks after cleaning or exposure to unusual feed constituents.

16.4.4 Compatibility of Dosing Chemicals:

Ensure that chemical dosing tanks, pumps, tubing, and injection points are made of materials compatible with the concentrated chemical being dosed (e.g., use PVDF or PP for strong acids/bases). Also, verify that different chemicals being dosed (e.g., antiscalant and SBS) are compatible with each other and won't react or precipitate if mixed accidentally in lines or day tanks.

16.5 Chapter Summary and Key Takeaways

The RO concentration process creates aggressive chemical conditions requiring careful consideration of scaling, corrosion, and material compatibility.

- **Scaling Chemistry:** Driven by supersaturation. $CaCO_3$ (pH-dependent, LSI predicts), $CaSO_4$ (solubility limit), SiO_2 (polymerization, pH-dependent), and trace scales ($BaSO_4$, $SrSO_4$) are key concerns. Antiscalants interfere with crystal nucleation and growth.
- **Corrosion:** Threatens metallic components. Pitting/crevice corrosion driven by chlorides is the major concern for stainless steels. Proper material selection (based on Cl^-, temp) and water chemistry control are key mitigation strategies.
- **Chemical Compatibility:** All materials must be compatible with feed/concentrate water and treatment/cleaning chemicals.
    - **TFC Membranes:** Intolerant of oxidants (chlorine); strict pH and temperature limits apply.
    - **Piping/Vessels:** Correct stainless steel grade or appropriate plastic is crucial.
    - **Elastomers (Seals):** Material (e.g., EPDM, Viton®) must match water chemistry and cleaning agents to prevent leaks.
    - **Dosing Systems:** Must be compatible with concentrated chemicals.

A holistic understanding of these chemical interactions is vital for designing robust systems, selecting appropriate materials, preventing premature failures, and optimizing long-term operation.

16.6 References

1. Stumm, W., & Morgan, J. J. (1996). Aquatic Chemistry: Chemical Equilibria and Rates in Natural Waters (3rd ed.). Wiley-Interscience. (Fundamental text on water chemistry and mineral equilibria).

2. Nalco Water. The Nalco Water Handbook (Multiple Editions). McGraw-Hill. (Practical sections on scaling and corrosion control in water systems).
3. Jones, D. A. (1996). Principles and Prevention of Corrosion (2nd ed.). Prentice Hall. (Comprehensive textbook on corrosion science and engineering).
4. Schweitzer, P. A. (Ed.). (2006). Corrosion Resistance Tables: Metals, Plastics, Nonmetallics, and Rubbers (5th ed.). CRC Press. (Extensive compatibility data).
5. Antiscalant and Membrane Manufacturer Technical Literature: Provide crucial data on scaling limits with specific antiscalants and chemical compatibility constraints for membranes and elastomers.
6. Amjad, Z. (Ed.). (1993). Reverse Osmosis—Membrane Technology, Water Chemistry, and Industrial Applications. Van Nostrand Reinhold. (Contains detailed chapters on water chemistry and scaling relevant to RO).
7. Iler, R. K. (1979). The Chemistry of Silica: Solubility, Polymerization, Colloid and Surface Properties, and Biochemistry. Wiley. (Definitive, though highly technical, reference on silica chemistry).

Chapter 17: Emerging Trends in RO Technology - Peeking Over the Horizon

17.1 Introduction: The Continuous Evolution of RO

Reverse Osmosis has transformed water treatment over the past half-century, evolving from a laboratory curiosity to a global cornerstone technology. Yet, the journey is far from over. The fundamental challenges remain: reducing the energy required for separation (especially for seawater), pushing recovery rates higher to conserve water and minimize brine volume, developing membranes more resistant to fouling, improving the removal of specific challenging contaminants, and lowering overall costs.

Driven by these persistent needs and enabled by advances in materials science, nanotechnology, process engineering, and data science, RO technology continues to evolve. Researchers and engineers worldwide are actively exploring innovative materials, novel process configurations, and smarter operating strategies. This chapter provides a glimpse into some of the key emerging trends that promise to enhance the performance, efficiency, and sustainability of Reverse Osmosis in the coming years and decades. While some are closer to commercial reality than others, they collectively point towards a future where RO plays an even more significant role in addressing global water challenges.

17.2 Pushing the Boundaries: Advanced Membrane Materials

The heart of RO is the membrane, and much research focuses on creating materials with superior permeability (letting water through easily), selectivity (blocking salts effectively), and durability (resisting fouling, chemicals, and pressure).

17.2.1 Biomimetic Membranes: Learning from Nature (Aquaporins)

Concept: Biological cell membranes transport water with remarkable efficiency and selectivity using specialized protein channels called aquaporins (AQPs). These channels allow water molecules to pass through single file at extremely high rates while excluding ions and other solutes. Biomimetic research aims to incorporate these natural AQP proteins into synthetic membrane structures or create synthetic channels that mimic their function.

Potential: Ultra-high water permeability at lower pressures, potentially leading to significant energy savings. Excellent selectivity due to the precisely structured channel.

Challenges: Stability and longevity of embedded proteins in industrial conditions, cost-effective large-scale manufacturing, maintaining the delicate protein structure during membrane fabrication and operation. Commercial examples exist but are still finding their niche.

17.2.2 Thin-Film Nanocomposite (TFN) Membranes Revisited

Concept: While mentioned in Chapter 4, the incorporation of nanoparticles into the traditional polyamide TFC structure remains a major area of active research and development, moving towards commercialization. Various nanoparticles (zeolites, MOFs, CNTs, graphene oxide, TiO_2, silica) are embedded within the polyamide active layer during or after interfacial polymerization.

Potential: Nanoparticles can potentially enhance water permeability (by creating preferential pathways or modifying polymer structure), improve selectivity, increase mechanical strength, enhance chlorine resistance (e.g., protective effect), or impart anti-fouling/anti-microbial properties (e.g., TiO_2 photocatalysis, Ag nanoparticles).

Challenges: Achieving uniform dispersion of nanoparticles without aggregation, ensuring long-term stability and adhesion within the polymer matrix, potential leaching of nanoparticles, scalability, and cost-effectiveness. Performance improvements are often incremental rather than revolutionary so far.

17.2.3 Carbon-Based Materials: Graphene and Carbon Nanotubes (CNTs)

Concept: Utilizing the unique properties of carbon nanomaterials.

- **Graphene/Graphene Oxide (GO):** Atomically thin sheets of carbon atoms. Precise sub-nanometer pores can theoretically be created in graphene sheets for highly selective molecular sieving. GO, with its oxygen functional groups, can be incorporated into TFNs or used to form laminar membrane structures.
- **Carbon Nanotubes (CNTs):** Hollow cylinders of graphene. Can act as extremely smooth, low-friction channels for rapid water transport when vertically aligned within a membrane matrix.

Potential: Extremely high water flux due to the low friction and confinement effects within these structures. Potentially high selectivity if pore sizes can be precisely controlled (especially for pure graphene).

Challenges: Fabricating large-area, defect-free graphene membranes remains a major hurdle. Aligning CNTs vertically on a large scale is difficult and costly. Controlling pore size distribution and ensuring stability are ongoing issues. Integration into scalable module formats is complex. Still largely in the research phase for RO applications.

17.2.4 Metal-Organic Frameworks (MOFs) and Covalent Organic Frameworks (COFs)

Concept: Crystalline porous materials with highly ordered structures built from metal ions or clusters linked by organic ligands (MOFs) or purely organic building blocks linked by strong covalent bonds (COFs). They offer tunable pore sizes (often sub-nanometer), extremely high surface areas, and potentially tailored chemical functionalities within the pores.

Potential: Can be used as nanoparticles in TFN membranes (see 17.2.2) or potentially formed as continuous thin films. Their well-defined pores could offer high permeability and precise molecular sieving capabilities for enhanced selectivity.

Challenges: Stability of some MOFs in water (hydrolytic stability) can be an issue. Fabricating continuous, defect-free thin films of MOFs/COFs on suitable supports is difficult. Scalability and cost remain concerns.

17.2.5 Towards Enhanced Selectivity: Ion-Specific Membranes

Concept: Moving beyond simply rejecting all salts, research aims to develop membranes that can selectively remove specific "target" ions while allowing other desirable or harmless ions to pass. This could involve incorporating specific ion-binding sites or channels into the membrane structure or utilizing specialized electro-driven membrane processes.

Potential Applications: Removing specific toxic ions (e.g., arsenic, nitrate) without removing beneficial minerals, selectively removing hardness ions (softening), recovering valuable ions (e.g., lithium, magnesium) from brine streams, or tailoring water composition for specific industrial uses.

Challenges: Achieving high selectivity between ions of similar size and charge is chemically difficult. Maintaining high water flux alongside high selectivity is challenging. Still largely in the research and early development stage for practical RO-like applications.

17.3 Beyond Material Science: Nanotechnology Integration Strategies

Nanotechnology isn't just about new bulk materials; it's also about manipulating surfaces and structures at the nanoscale to improve performance, particularly regarding fouling.

17.3.1 Surface Patterning and Nanostructuring for Fouling Control:

Concept: Creating specific nanoscale topographical patterns (ridges, valleys, bumps) or chemical patterns on the membrane surface. Inspired by natural surfaces like lotus leaves (superhydrophobicity) or shark skin (reduced microbial adhesion).

Potential: These patterns can potentially reduce the contact area available for foulants, alter local hydrodynamics to hinder attachment, or create surfaces that are inherently less "sticky" to specific foulants like bacteria or proteins.

Challenges: Creating precise, durable nanoscale patterns uniformly over large membrane areas cost-effectively. Demonstrating long-term effectiveness under real-world operating conditions.

17.3.2 Catalytic and Reactive Nanoparticles for Self-Cleaning:

Concept: Incorporating nanoparticles into the membrane surface or coating that can actively degrade foulants when triggered (e.g., photocatalytic TiO_2 nanoparticles activated by UV light to break down organics) or release antimicrobial agents slowly over time.

Potential: Reduce the need for frequent chemical cleaning by enabling in-situ fouling degradation or prevention, leading to more stable operation and potentially longer membrane life.

Challenges: Ensuring efficient activation (e.g., UV penetration), long-term stability and activity of the nanoparticles, preventing nanoparticle leaching, potential formation of unwanted byproducts, cost.

17.4 The Quest for Efficiency: Energy-Efficient RO Systems

Reducing the energy footprint of RO, especially SWRO, remains a primary driver of innovation.

17.4.1 Continued Improvements in Low-Energy Membranes:

Manufacturers continue to refine TFC polyamide membranes to achieve higher water permeability (higher 'A' value), allowing operation at lower feed pressures for the same flux, particularly for BWRO and water reuse applications.

17.4.2 Advanced Energy Recovery Devices (ERDs):

While ERDs (like pressure exchangers and turbine-based devices) are already standard in SWRO, research continues on improving their efficiency (approaching 98-99%), reducing their footprint and cost, and enhancing their reliability and operational flexibility.

17.4.3 Process Optimization and Smart Controls (AI/ML):

Concept: Using advanced process modeling, sensor networks, and Artificial Intelligence (AI) / Machine Learning (ML) algorithms to optimize RO operation in real-time.

Potential: AI could predict fouling onset more accurately, optimize cleaning schedules, adjust operating parameters (pressure, flow, recovery) dynamically based on feed quality and energy price fluctuations to minimize overall costs, detect anomalies indicative of leaks or failures, and potentially optimize ERD operation.

Challenges: Requires robust sensor networks, high-quality data, development of accurate predictive models, and integration with existing control systems (PLCs/SCADA).

17.4.4 Renewable Energy Integration:

Coupling RO plants (especially SWRO) directly with renewable energy sources like solar PV or wind power is crucial for sustainable desalination. This involves managing the intermittent nature of renewables, potentially using battery storage or flexible RO operation strategies.

17.5 Thinking Outside the Box: Forward Osmosis and Hybrid Systems

While RO uses hydraulic pressure, other membrane processes utilize different driving forces. Combining these processes holds promise for specialized applications or improved efficiency.

17.5.1 Forward Osmosis (FO): Osmotic Pressure Driven Separation

17.5.1.1 The FO Concept and Draw Solutions:

FO uses a semi-permeable membrane similar to RO, but the driving force is the natural osmotic pressure difference between a relatively dilute feed solution (e.g., wastewater) and a highly concentrated "draw" solution on the other side of the membrane. Water naturally moves from the feed into the draw solution, leaving contaminants behind. No high hydraulic pressure is applied.

17.5.1.2 Potential Applications:

Treating highly fouling feed streams (lower fouling tendency due to lack of hydraulic pressure), concentrating wastewater streams with minimal energy for the separation step itself, potentially as a pretreatment for RO desalination (by drawing water from seawater into a recoverable draw solution).

17.5.1.3 Challenges:

The major hurdle is the draw solution regeneration step. After water is drawn into the draw solution, the water must be separated from the draw solute to produce the final product. This regeneration step often requires significant energy (e.g., using RO or thermal processes to reconcentrate the draw solution), potentially offsetting the energy savings of the FO step itself. Development of efficient, low-energy draw solutes and regeneration methods, along with high-flux, low-fouling FO membranes, is critical for wider adoption.

17.5.2 Hybrid Systems: Combining Strengths

Integrating RO with other membrane or separation processes can leverage the advantages of each technology:

- **RO-MD (Membrane Distillation):** MD uses a hydrophobic membrane and a temperature difference to evaporate water from a feed stream and condense it on the permeate side. It can treat very high salinity brines (where RO osmotic pressure is too high) and is driven by low-grade heat rather than high pressure. Hybrid RO-MD systems could potentially achieve very high overall recovery by using MD to treat the RO concentrate. MD faces challenges with scaling, membrane wetting, and energy efficiency compared to optimized RO within RO's operating range.
- **RO-ED (Electrodialysis):** ED uses ion exchange membranes and an electric field to remove salts. It becomes more energy-efficient than RO at lower salinities. Hybrid systems might use ED to polish RO permeate or potentially treat certain streams where selective ion removal is beneficial.
- **FO-RO Combinations:** As mentioned, FO could pretreat challenging water before RO, or RO could be used to regenerate the draw solution in an FO process.

17.6 Tackling the Waste: Advanced Brine Management and Resource Recovery

As RO use expands, managing the large volumes of salty concentrate (brine) produced becomes an increasingly critical environmental and economic challenge, especially for inland plants.

17.6.1 High-Recovery RO Designs:

Pushing the limits of RO recovery itself through optimized staging, interstage treatment, or specialized membranes reduces the initial brine volume.

16.6.2 Brine Concentration Technologies:

Technologies specifically designed to further concentrate the RO brine, minimizing the final volume for disposal:

- **Osmotically Assisted RO (OARO) / Counter-Flow RO (CFRO):** Use clever flow configurations or osmotic agents to overcome the high osmotic pressure of the brine, allowing further water extraction via RO membranes beyond conventional limits.
- **Membrane Distillation (MD) / Forward Osmosis (FO):** Can potentially concentrate brines beyond RO's pressure limits, as discussed above.
- **Electrodialysis (ED/EDR):** Can be used for brine concentration in some contexts.
- **Thermal Evaporators/Crystallizers:** Traditional but energy-intensive methods used as the final step in ZLD systems to produce solid salt waste. Reducing the feed volume to these units via membrane pre-concentration is key to lowering ZLD costs.

17.6.3 Selective Mineral Extraction from Brine:

Concept: Viewing RO brine not just as waste, but as a potential source of valuable minerals like magnesium, calcium, potassium, lithium, bromine, etc.

Technologies: Developing selective precipitation techniques, advanced ion exchange resins, or specialized electro-membrane processes to selectively extract target minerals from the complex brine matrix.

Potential: Create valuable byproducts, offset desalination costs, reduce environmental impact of brine disposal, contribute to circular economy principles. This is an active area of research and pilot testing.

17.7 Chapter Summary and Key Takeaways

- RO technology is continuously evolving, driven by needs for lower energy, higher recovery, better fouling resistance, and enhanced selectivity.
- Advanced membrane materials are a key focus, including biomimetic (aquaporin) membranes, thin-film nanocomposites (TFNs), carbon-based materials (graphene, CNTs), and MOFs/COFs, all aiming for higher permeability and/or selectivity.
- Nanotechnology is also applied via surface patterning or reactive coatings to mitigate fouling.
- Energy efficiency improvements continue through better membranes, advanced ERDs, smart controls (AI/ML), and integration with renewable energy.
- Forward Osmosis (FO) offers potential for low-fouling treatment and concentration but faces challenges with draw solution regeneration. Hybrid systems combining RO with MD, ED, or FO aim to leverage complementary strengths.
- Advanced brine management is crucial, involving high-recovery RO designs, brine concentration technologies (OARO, MD, etc.), and the emerging field of selective mineral recovery from brine to create value from waste.
- While challenges remain in scaling up and commercializing many of these trends, they indicate a vibrant future for membrane-based water treatment with improved performance and sustainability.

17.8 References

1. Elimelech, M., & Phillip, W. A. (2011). The Future of Seawater Desalination: Energy, Technology, and the Environment. Science, 333(6043), 712–717. (Excellent overview of future directions).
2. Werber, J. R., Osuji, C. O., & Elimelech, M. (2016). Materials for next-generation desalination and water purification membranes. Nature Reviews Materials, 1(5), 16018. (Focuses on advanced materials).

3. Zhao, D. L., Japip, S., Zhang, Y., Weber, M., Maletzko, C., & Chung, T. S. (2020). Emerging Forward Osmosis (FO) Technologies and Challenges: A Comprehensive Review. Water Research, 173, 115641.
4. Karim, M. R., Mansur, D., Kurnia, K. A., & Pasek, A. D. (2019). Metal–organic frameworks (MOFs) for aqueous separations. Current Opinion in Chemical Engineering, 23, 122-130.
5. Cohen-Tanugi, D., & Grossman, J. C. (2012). Water desalination across nanoporous graphene. Nano letters, 12(7), 3602-3608. (Example of graphene research).
6. Tong, T., & Elimelech, M. (2016). The global rise of zero liquid discharge for wastewater management: Drivers, technologies, and future directions. Environmental Science & Technology, 50(13), 6846–6855.
7. Logan, B. E., & Elimelech, M. (2012). Membrane-based processes for sustainable power generation using water. Nature, 488(7411), 313-319. (Discusses related membrane energy technologies).
8. Aquaporin A/S, Forward Osmosis Technologies manufacturers (e.g., Hydration Technology Innovations - HTI), Brine concentration technology providers (e.g., Gradiant Corp). (Company literature often highlights emerging technology applications).

Chapter 18: Frequently Asked Questions (FAQs) - Practical Answers to Common Queries

18.1 Introduction: Answering Your RO Questions

Throughout this book, we've covered the principles, design, operation, and applications of Reverse Osmosis in considerable detail. However, practical questions often arise that benefit from direct, concise answers. This chapter serves that purpose, addressing frequently asked questions based on common points of interest or confusion encountered by operators, engineers, technicians, and students engaging with RO technology.

18.2 Basic Principles & Terminology

Q1: What's the difference between osmosis and reverse osmosis again, in simple terms?

A1: Think of osmosis as nature's way of balancing things. If you have salty water and pure water separated by a special filter (a semi-permeable membrane), water naturally flows from the pure side to the salty side to try and dilute it. This natural flow creates osmotic pressure. Reverse Osmosis (RO) flips this around. We apply hydraulic pressure (from a pump) to the salty side – pressure that's stronger than the natural osmotic pressure. This forces water molecules to go in the reverse direction, from the salty side through the membrane to the pure side, leaving the salts and other impurities behind. We're essentially overpowering nature's tendency. (See Chapter 2)

Q2: Why does RO need such high pressure?

A2: It needs enough pressure to do two things: first, overcome the natural osmotic pressure (π) of the feed water, which acts like a back-pressure resisting the flow; and second, provide the additional Net

Driving Pressure (NDP) needed to actually push water molecules through the tight structure of the RO membrane at a reasonable rate (flux). The saltier the water (higher TDS), the higher the osmotic pressure, and therefore, the higher the feed pressure required. Seawater RO needs much higher pressure (800-1200 psi) than brackish water RO (150-400 psi). (See Chapter 2 & 7)

Q3: Does RO remove everything from water? What does it not remove well?

A3: RO is highly effective at removing the vast majority (>95-99%+) of dissolved salts, minerals, larger organic molecules, bacteria, viruses, and suspended particles. However, it's not perfect. Things it generally doesn't remove well include:

- **Dissolved Gases:** Small, uncharged molecules like dissolved oxygen (O_2) and carbon dioxide (CO_2) pass readily through the membrane. CO_2 passage lowers permeate pH.
- **Some Small, Uncharged Organics:** Certain low molecular weight organic compounds (like some solvents or disinfection byproducts) may pass through to some extent.
- **Some Weakly Ionized Species (at neutral pH):** Species like silica (as silicic acid) and boron (as boric acid) are less effectively rejected at neutral pH compared to highly charged ions. Rejection improves if pH is adjusted to ionize them. (See Chapter 4 & 15)

Q4: What is TDS and why is it so important for RO?

A4: TDS stands for Total Dissolved Solids. It's the measure of all the inorganic salts (like sodium chloride, calcium sulfate) and small amounts of organic matter dissolved in the water, usually expressed in mg/L or ppm. It's crucial for RO because:

- It dictates the feed water's osmotic pressure, which determines the required operating pressure and energy consumption.
- It defines the challenge for the RO system – higher feed TDS means more salts to remove to achieve the desired permeate quality.

- It's often used as the primary indicator of permeate quality. (See Chapter 3 & 15)

Q5: What's the difference between RO, NF, UF, and MF?

A5: They are all membrane filtration processes, but they differ mainly in the size of contaminants they remove, the tightness of their membranes, and the pressure they require:

- Microfiltration (MF): Largest pores (approx. 0.1-1.0 μm). Removes suspended solids, bacteria, protozoa. Low pressure.
- Ultrafiltration (UF): Smaller pores (approx. 0.01-0.1 μm). Removes everything MF does, plus viruses, colloids, and larger organic molecules. Low to moderate pressure. Often used as RO pretreatment.
- Nanofiltration (NF): Tighter pores than UF (approx. 0.001-0.01 μm or 1-10 nm). Rejects larger dissolved molecules and multivalent ions (like Ca^{2+}, Mg^{2+} - providing softening), but passes more monovalent ions (like Na^+, Cl^-) than RO. Moderate pressure.
- Reverse Osmosis (RO): Tightest membrane structure (effectively non-porous according to solution-diffusion model). Rejects nearly all dissolved salts (monovalent and multivalent ions) and larger molecules. Requires high pressure to overcome osmotic pressure. (See Chapters 2, 4, 9)

18.3 Membranes and Modules

Q6: How long do RO membranes typically last?

A6: With proper pretreatment, appropriate system design, diligent operation, and timely cleaning, RO membranes (specifically TFC polyamide elements) typically last 3 to 7 years, sometimes longer in favorable conditions. Factors shortening life include frequent or harsh cleanings, irreversible fouling or scaling, physical damage, or chemical attack (especially by oxidants). Gradual, irreversible compaction or hydrolysis also limits lifespan.

Q7: Why are TFC polyamide membranes so sensitive to chlorine?

A7: The polyamide polymer structure contains amide functional groups (-NH-CO-). Chlorine (and other strong oxidants) readily attacks and breaks these chemical bonds. This damage disrupts the integrity of the thin rejection layer, causing a rapid and irreversible loss of salt rejection capability. The membrane essentially becomes leaky. This is why complete dechlorination before the RO is mandatory for TFC membranes. (See Chapter 4 & 16)

Q8: What's the difference between brackish water and seawater membranes?

A8: Both are typically TFC polyamide, but they are optimized differently:

- **Seawater (SW) Membranes:** Designed for maximum salt rejection (often >99.5%, sometimes >99.8%) to produce potable water from very high TDS feed. They generally have slightly lower water permeability ('A' value) and require higher operating pressures (800-1200 psi).
- **Brackish Water (BW) Membranes:** Designed for good salt rejection (e.g., 98-99.5%) from lower TDS feed. They typically have higher water permeability, allowing operation at lower pressures (150-400 psi) for energy efficiency. Low energy BW membranes push this even further.

Q9: What is a spiral-wound element? How does water flow through it?

A9: It's the most common way RO membrane sheets are packaged. Imagine layers of membrane sheets (active side out), permeate collection material (like fabric), and feed channel spacers (like plastic mesh) all sandwiched together and rolled up like a jelly roll around a central collection tube.

- **Feed/Concentrate Flow:** Pressurized feed water flows axially (along the length of the roll) through the channels created by the feed spacer mesh, parallel to the membrane surface. Water that doesn't permeate exits the other end as concentrate.

- **Permeate Flow:** Water passing through the membrane enters the permeate collection material, then flows spirally inward within that layer towards the central perforated tube, where it collects and exits the element. (See Chapter 4)

18.4 System Design and Operation

Q10: What does "recovery rate" mean and why can't it be 100%?

A10: Recovery Rate is the percentage of the incoming feed water that is turned into purified permeate water (Recovery % = Permeate Flow / Feed Flow * 100%). In cross-flow filtration like RO, you can't achieve 100% recovery because some water must continuously flow across the membrane surface (as concentrate) to sweep away the rejected salts and prevent them from building up excessively (concentration polarization) and causing severe scaling or fouling. The maximum achievable recovery is limited by the concentration at which salts in the reject stream start to precipitate (scaling). (See Chapter 3 & 7)

Q11: What is flux and why is it important to control?

A11: Flux is the rate at which permeate flows through a unit area of membrane surface (e.g., Gallons per Square Foot per Day - GFD). Operating at too high a flux forces water through the membrane too quickly. This causes salts to build up rapidly at the membrane surface (high concentration polarization), increasing the risk of scaling and fouling, and potentially reducing salt rejection. Designers choose a safe average flux based on water quality and membrane type, and operators monitor normalized flux to detect fouling. (See Chapter 3 & 7)

Q12: What does a 2:1 array mean? Why stage systems?

A12: A 2:1 array means the RO system has two stages: the first stage has 2 pressure vessels operating in parallel, and their combined concentrate stream feeds 1 pressure vessel in the second stage. Staging (arranging vessels in series like this) is necessary because as permeate is removed, the feed flow decreases. By reducing the number of parallel vessels in subsequent stages, we maintain adequate flow velocity (cross-flow) through the remaining feed channels, which is crucial for controlling concentration polarization and fouling, especially at higher recovery rates. (See Chapter 5)

Q13: What is a permeate flush and why is it critical?

A13: A permeate flush is a low-pressure rinse of the membrane elements using permeate water (or sometimes pretreated feed water), performed just before shutting down the RO system for more than a few hours. Its purpose is to displace the highly concentrated feed/concentrate water trapped inside the elements. Leaving concentrated water stagnant greatly increases the risk of scaling as water cools, and provides food for microbial growth. Regularly performing this flush is one of the most important operational steps to protect membranes during shutdowns. (See Chapter 10)

Q14: How important is feed water temperature?

A14: Very important! Temperature significantly affects membrane performance:

- **Flux:** Higher temperature decreases water viscosity, making water pass through membranes more easily, thus increasing permeate flow (flux). Lower temperature decreases flow.
- **Salt Passage:** Higher temperature slightly increases the rate at which salts diffuse through the membrane, potentially worsening permeate quality slightly.
- **Scaling:** Affects solubility limits ($CaCO_3$ less soluble, $CaSO_4$/SiO_2 more soluble at higher temps).
- **Normalization:** Because of these strong effects, performance data MUST be normalized for temperature to accurately assess fouling trends. (See Chapters 2, 10, 15)

18.5 Pretreatment Concerns

Q15: Why is pretreatment so important? What happens if it's inadequate?

A15: Pretreatment prepares the raw water to be suitable for the sensitive RO membranes. It removes suspended solids, colloids, potentially harmful chemicals (like chlorine), microbes, and adjusts chemistry to control scaling. Inadequate pretreatment is the single most common cause of RO problems. It leads directly to rapid membrane fouling and/or scaling, resulting in low flow, high pressure drop, poor water quality, frequent need for aggressive cleaning (which shortens membrane life), and potentially irreversible membrane damage. Investing in proper pretreatment saves money in the long run. (See Chapters 8 & 9)

Q16: What is SDI and how often should it be measured?

A16: SDI (Silt Density Index) is an empirical test measuring the plugging potential of feed water due to fine suspended solids and colloids (the main physical foulants). It involves timing how quickly a 0.45-micron filter plugs under standard pressure. The target for RO feed is typically SDI < 3 (definitely < 5). It should be measured after the final pretreatment step (e.g., after cartridge filters) regularly – daily or even per shift, especially if feed water quality is variable. It's your primary indicator of particulate fouling risk. (See Chapter 3 & 8)

Q17: What does an antiscalant do? Is it always needed?

A17: An antiscalant is a chemical injected into the feed water to prevent mineral scale (like calcium carbonate, calcium sulfate) from forming on the membranes, even when the water becomes supersaturated in the concentrate stream. It interferes with crystal formation and growth. Whether it's needed depends on the feed water chemistry (hardness, alkalinity, sulfate, etc.) and the target recovery rate. For many brackish water systems operating at moderate to high recovery (>50-60%), an antiscalant is essential. For seawater or very low recovery systems, it might sometimes be omitted, but scaling potential must always be carefully calculated using design software. (See Chapter 9 & 16)

18.6 Cleaning and Maintenance

Q18: How often should RO membranes be cleaned?

A18: There's no fixed schedule. Cleaning frequency depends entirely on the rate of fouling, which is determined by feed water quality and pretreatment effectiveness. Cleaning should be performed proactively based on performance monitoring. The trigger is typically when normalized permeate flow drops by 10-15%, or normalized differential pressure increases by 15-20%, or normalized salt passage increases by 10-15%. This could be anywhere from every few weeks to once a year or less in well-managed systems with good feed water. (See Chapter 11)

Q19: What happens if I wait too long to clean?

A19: Delaying cleaning allows fouling layers to become thicker, more compacted, and potentially less chemically reactive (especially biofilms). This makes cleaning much less effective, requiring longer cleaning times, potentially more aggressive chemicals (risking membrane damage), and often results in incomplete performance recovery. Severe, neglected fouling can cause irreversible flux loss or physical damage to the elements. Clean early, clean effectively! (See Chapter 11)

18.7 Water Quality and Applications

Q20: Is RO water safe to drink? Does it need minerals added back?

A20: RO removes contaminants very effectively, making the water microbiologically safe and removing harmful chemicals. However, it also removes beneficial minerals like calcium and magnesium. This demineralized water can taste flat and be slightly corrosive. For municipal drinking water or home use, remineralization (adding back calcium and magnesium carbonates) is often practiced after RO. This improves taste, makes the water less corrosive to pipes, and restores some mineral content. Even without remineralization, the water is safe in terms of regulated contaminants, but may not be ideal for taste or plumbing. (See Chapter 14 & 15)

Q21: Can RO remove viruses and bacteria?

A21: Yes. The physical structure of RO membranes is much tighter than the size of bacteria and viruses. RO provides an excellent physical barrier, achieving very high removal rates (> 99.99%+ or log removal values > 4). It's considered a key part of a multi-barrier approach for ensuring microbial safety in drinking water and high-purity applications. (See Chapter 8 & 15)

Q22: What happens to the concentrate (reject water)?

A22: This depends on the location and regulations.

- **Coastal Plants:** Concentrate (brine, especially from SWRO) is usually discharged back to the sea, often via diffusers designed to promote rapid mixing and minimize localized salinity increases.
- **Inland Plants:** Concentrate disposal is more challenging. Options include discharge to a sewer (if permitted by the municipality, may require TDS limits), deep well injection, evaporation ponds (requires large land area and suitable climate), or treatment in brine concentrators/crystallizers as part of a Minimal or Zero Liquid Discharge (MLD/ZLD) system (expensive). (See Chapter 13 & 17)

18.8 Conclusion: Continuous Learning

These questions cover some common ground, but every RO system and water source is unique. The key is to understand the fundamental principles laid out in this book, operate your system diligently, monitor performance carefully using normalized data, and approach troubleshooting systematically. Don't hesitate to consult manufacturer manuals, technical resources, and experienced colleagues. The field of water treatment is one of continuous learning, and mastering Reverse Osmosis is a rewarding part of that journey.

Appendices

Appendix A: Unit Equivalents and Conversions

(Based on units and conversions mentioned or implied in Chapters 1-18)

Unit / Concept	Equivalent / Conversion / Note
Pressure	
psi	Pounds per Square Inch
atm	Atmosphere (≈ 14.7 psi)
bar	Bar (≈ 14.5 psi ≈ 100 kPa)
kPa	Kilopascal (≈ 0.145 psi)
psid	psi differential (pressure drop)
Flow Rate	
GPD	Gallons Per Day
GPM	Gallons Per Minute
m^3/hr (or m^3/h)	Cubic Meters per Hour
m^3/day	Cubic Meters per Day
LPM (or L/min)	Liters Per Minute
L/hr	Liters per Hour
Flux (Flow per Unit Area)	
GFD	Gallons per Square Foot per Day (membrane area)
LMH (or $L/m^2 \cdot h$)	Liters per Square Meter per Hour (1 GFD ≈ 1.7 LMH)
Concentration	
mg/L	Milligrams per Liter (≈ ppm for dilute aqueous solutions)
ppm	Parts Per Million (≈ mg/L for dilute aqueous solutions)

ppb	Parts Per Billion (μg/L)
ppt	Parts Per Trillion (ng/L)
μg/L	Micrograms per Liter (= ppb)
mol/L	Moles per Liter (Molar concentration)
mg/L as $CaCO_3$	Equivalent concentration for Hardness & Alkalinity
Conductivity	
μS/cm	Microsiemens per centimeter
mS/cm	Millisiemens per centimeter
TDS / Conductivity Factor	TDS (mg/L) ≈ Conductivity (μS/cm) * (0.5 - 0.8)
Temperature	
°C	Degrees Celsius
°F	Degrees Fahrenheit (°F = °C * 9/5 + 32)
K	Kelvin (K = °C + 273.15)
Size / Length	
μm	Micrometer (micron) (10^{-6} meters)
nm	Nanometer (10^{-9} meters)
mil	Thousandth of an inch (used for spacer thickness)
Other	
NTU	Nephelometric Turbidity Unit
mV	Millivolt (used for ORP)
CFU/mL	Colony Forming Units per Milliliter (for bacteria counts)
pH	Measure of acidity/basicity (log scale)
ORP	Oxidation-Reduction Potential (mV)
R (Ideal Gas Constant)	0.08206 L·atm/(mol·K) or 8.314 L·kPa/(mol·K) (Used in Van't Hoff Eq.)

Appendix B: Glossary

(Based on terms defined or emphasized in Chapters 1-18)

Term (and Abbreviation)	Definition/Description (based on book content)
Alkalinity	Water's acid-neutralizing capacity, primarily due to bicarbonate/carbonate ions. Crucial for $CaCO_3$ scaling prediction (LSI). Measured as mg/L as $CaCO_3$.
Antiscalant	A chemical injected into feed water to inhibit mineral scale formation on membranes, allowing operation under supersaturated conditions. Works via threshold inhibition, crystal modification, dispersion.
Aquaporin (AQP)	Protein channels in biological membranes facilitating highly efficient and selective water transport. Inspiration for biomimetic membranes.
Array	The arrangement of pressure vessels into stages in an RO system, denoted by the number of vessels per stage (e.g., 2:1, 4:2:1).
Asymmetric Membrane	A membrane structure (like early CA membranes) with a thin, dense "skin" layer for separation supported by a thicker, porous substructure.
Autopsy (Membrane)	Destructive analysis of a fouled or failed membrane element performed by a specialized lab to identify foulants, scaling, or damage definitively.
Backwashing	Reversing flow through a media filter (or MF/UF) to fluidize the bed and flush out trapped solids to waste.
Beta (β) / Concentration Factor (CF)	The ratio of solute concentration in the concentrate stream to the feed stream ($\beta \approx 1 / (1 - \text{Recovery})$). Indicates how concentrated the reject becomes, crucial for scaling predictions.
Biofilm	A slimy layer composed of microorganisms (bacteria, etc.) embedded in an Extracellular Polymeric Substance (EPS) matrix, adhering to surfaces like membranes and spacers.

Biofouling	Fouling caused by the growth of microorganisms and formation of biofilm on membrane surfaces/spacers, typically causing high ΔP and flux loss.
Biomimetic Membrane	A membrane designed to mimic biological transport mechanisms, often incorporating or simulating aquaporins for high water permeability and selectivity.
Blowdown (Cooling Tower)	Water intentionally drained from a cooling tower system to control the concentration of dissolved solids. RO can be used to treat blowdown for reuse.
Boiler Feedwater (BFW)	High-purity water required for feeding boilers (especially high-pressure ones) to prevent scale and corrosion. RO is a key technology for BFW production.
Boron (B)	An element present in seawater that is poorly rejected by standard RO at neutral pH and can be toxic to plants at low concentrations. Removal requires special membranes or high pH second pass.
Brackish Water (BW)	Water with moderate salinity (TDS typically 1,000-15,000 mg/L), less salty than seawater but saltier than freshwater.
Brine	A general term for highly concentrated salt solutions, often used interchangeably with RO concentrate, especially from SWRO.
Cartridge Filter	A final polishing/safety filter (typically 5-micron nominal) installed just before the high-pressure pump to protect it and the RO membranes from stray particulates. Operates in dead-end mode.
Cavitation	Formation and collapse of vapor bubbles in a pump suction due to insufficient inlet pressure (NPSH), causing noise, vibration, and severe damage.
Cellulose Acetate (CA)	The first polymer material used to make practical asymmetric RO membranes. Has some chlorine tolerance but limited pH range and lower performance than TFC.

Chelating Agent	A chemical (like EDTA) that binds metal ions into soluble complexes, used in cleaning solutions to help dissolve metal-based scales ($CaSO_4$, metal oxides).
Clean-In-Place (CIP)	The process of chemically cleaning RO membranes without removing them from the system, typically using a dedicated CIP skid with tanks, pumps, and heaters.
Coagulation	A pretreatment step using chemicals (e.g., alum, ferric salts, polymers) to destabilize and neutralize the charge of colloidal particles, allowing them to aggregate. Requires rapid mixing.
Colloid	Very fine particles (approx. 0.001-1 μm) that remain suspended in water due to size and surface charge, contributing significantly to RO fouling potential (SDI).
Compaction (Membrane)	Irreversible compression of the membrane polymer structure under prolonged high pressure, leading to a gradual decline in water flux over time.
Concentration Factor (CF) / Beta (β)	See Beta.
Concentration Polarization (CP)	The accumulation of rejected solutes in a thin boundary layer adjacent to the membrane surface, leading to higher local osmotic pressure, reduced NDP, increased salt passage, and enhanced scaling risk.
Concentrate (Reject / Brine)	The portion of feed water that does not pass through the membrane, containing the concentrated rejected salts and contaminants.
Conductivity	A measure of water's ability to conduct electricity, used as a quick proxy for TDS and the primary real-time indicator of permeate quality. Measured in μS/cm or mS/cm.
Corrosion	Degradation of metallic components (pumps, pipes, vessels) through electrochemical reactions with the

	water. Pitting (due to chlorides) is a major concern for stainless steel.
Cross-Flow Filtration (CFF)	Filtration mode used in RO where feed water flows parallel (tangential) to the membrane surface, sweeping away rejected species and minimizing fouling, producing separate permeate and concentrate streams.
Dead-End Filtration (DEF)	Filtration mode where all fluid flows perpendicularly through the filter medium (like a coffee filter or cartridge filter), trapping solids on the surface. Not suitable for RO membranes directly.
Dechlorination	The process of removing chlorine or other oxidizing disinfectants (harmful to TFC membranes) from the feed water, typically using Sodium Bisulfite (SBS) injection or Granular Activated Carbon (GAC).
Desalination	The process of removing salts from water, typically seawater or brackish water, to produce fresh water. RO is the dominant desalination technology.
Differential Pressure (ΔP)	The pressure drop occurring as water flows through a component (cartridge filter, membrane element, stage, vessel). Increasing normalized ΔP is a key indicator of fouling/plugging.
Dissolved Air Flotation (DAF)	A pretreatment process where microscopic air bubbles attach to flocculated particles, floating them to the surface for removal. Effective for low-density flocs.
Double Pass RO	A system configuration where the permeate from a first RO pass becomes the feed for a second RO pass, used to achieve very high purity water (e.g., UPW, boiler feed).
Electrodeionization (EDI)	A polishing technology often used after RO, employing ion exchange membranes, resins, and an electric field to remove residual ions without chemical regeneration, producing high-purity water.

Element (Membrane Element / Module)	The individual, replaceable unit containing the RO membrane, spacers, and permeate tube, typically in a spiral-wound or hollow fiber configuration. Loaded into pressure vessels.
Energy Recovery Device (ERD)	A device used primarily in SWRO to capture hydraulic energy from the high-pressure concentrate stream and transfer it to the incoming feed stream, significantly reducing net energy consumption.
Feed Spacer	A plastic mesh material placed between membrane leaves in a spiral-wound element to create channels for feed water flow and promote turbulence to control CP.
Feed Water	The raw or pretreated water supplied to the RO system inlet.
Fiberglass Reinforced Plastic (FRP)	A composite material commonly used for RO pressure vessels due to its high strength, corrosion resistance, and smooth surface.
Flocculation	The pretreatment step following coagulation where destabilized particles are gently mixed to promote their aggregation into larger, settleable/filterable flocs.
Flux (Membrane Flux / Water Flux, Jw)	The rate of permeate flow per unit of membrane area (e.g., GFD or LMH). A key performance and design parameter. Normalized flux decline indicates fouling.
Fouling	The accumulation of unwanted material (particulates, microbes, organics, scale) on the membrane surface or within feed channels, leading to performance decline (lower flux, higher ΔP, lower rejection).
Forward Osmosis (FO)	A membrane process driven by osmotic pressure difference (using a concentrated "draw" solution) rather than hydraulic pressure. Potential for low-fouling treatment or brine concentration.
Granular Activated Carbon (GAC)	Porous carbon material used in filter beds for adsorption of organics and dechlorination.

Hardness	Primarily the concentration of dissolved calcium (Ca^{2+}) and magnesium (Mg^{2+}) ions, major contributors to scale formation. Expressed as mg/L as $CaCO_3$.
Hollow Fine Fiber (HFF)	A membrane module configuration using thousands of fine, self-supporting membrane fibers bundled together. Offers very high packing density but susceptible to plugging.
Human-Machine Interface (HMI)	A graphical display (often touchscreen) on the RO control panel allowing operator interaction, monitoring, and control.
Hydraulic Pressure	Mechanical pressure applied by a pump to the feed water to overcome osmotic pressure and drive water through the membrane in RO.
Hydrolysis	Chemical breakdown of a polymer (like CA or polyamide) by reaction with water, often accelerated by extreme pH or high temperature. Leads to irreversible membrane degradation.
Interfacial Polymerization	The chemical reaction process used to form the ultrathin polyamide rejection layer on TFC membranes, occurring at the interface between an aqueous amine solution and an organic acyl chloride solution.
Langelier Saturation Index (LSI)	A calculated index predicting the tendency of water to precipitate or dissolve calcium carbonate ($CaCO_3$) based on pH, Ca^{2+}, alkalinity, TDS, and temperature. Used to assess scaling risk in concentrate.
Manifolding	The system of pipes and fittings on the RO skid that distributes feed water and collects permeate and concentrate streams.
Media Filtration	A pretreatment process using beds of granular material (sand, anthracite, garnet) to remove suspended solids. Common types include dual media and multimedia filters.

Membrane	The thin, semi-permeable barrier that performs the separation in RO, allowing water passage while rejecting dissolved solutes.
Microfiltration (MF)	Pressure-driven membrane process with pore sizes ~0.1-1.0 μm, removing suspended solids and bacteria. Often used as RO pretreatment.
Module (Membrane Module / Element)	See Element.
Nanofiltration (NF)	Membrane process tighter than UF but looser than RO, rejecting larger molecules and divalent ions while passing more monovalent ions. Operates at moderate pressure.
Natural Organic Matter (NOM)	Complex mixture of dissolved organic compounds (e.g., humic/fulvic acids) from natural sources, often causing color and organic fouling of RO membranes.
Net Driving Pressure (NDP)	The effective pressure difference across the membrane that drives water permeation (NDP = P_feed - P_permeate - $\Delta\pi$). Decreases along the flow path.
Normalization (Data Normalization)	Mathematical adjustment of operating data (flux, salt passage, ΔP) to a standard set of reference conditions (temp, pressure, etc.) to remove operational variations and reveal true performance trends due to fouling or degradation.
Osmosis	The natural movement of solvent (water) across a semi-permeable membrane from a region of lower solute concentration to a region of higher solute concentration.
Osmotic Pressure (π)	The pressure that needs to be applied to a solution to prevent the inward flow of pure solvent across a semi-permeable membrane. Proportional to solute concentration (TDS).

Oxidation-Reduction Potential (ORP)	A measure (in mV) of water's tendency to oxidize or reduce substances. Used to monitor the presence/absence of oxidants like chlorine (target < +150-200 mV before TFC membranes).
Permeability Coefficient (A value)	A measure of the membrane's intrinsic ability to pass water under a given driving pressure. Related to water flux (Jw = A * NDP).
Permeate (Product Water)	The purified water that has passed through the RO membrane.
Permeate Flush	A critical shutdown procedure where low-pressure permeate water is used to flush high-concentration feed/concentrate water out of the membrane elements to prevent scaling/biofouling during standstill.
pH	A measure of acidity/basicity (0-14 scale). Critically affects $CaCO_3$/silica scaling, membrane stability, and rejection of some species.
Pitting Corrosion	Localized corrosion of metals (esp. stainless steel by chlorides) that creates small holes or pits, potentially leading to leaks.
Polyamide (PA)	The polymer typically used to form the ultrathin rejection layer in modern Thin-Film Composite (TFC) RO membranes. Highly effective but sensitive to oxidants like chlorine.
Positive Displacement (PD) Pump	A type of pump (e.g., plunger pump) that delivers a fixed volume per stroke/rotation. Can achieve high pressures efficiently but produces pulsating flow. Sometimes used for RO feed.
Pretreatment	The series of treatment steps applied to raw water *before* it reaches the RO membranes to remove harmful contaminants (solids, colloids, organics, microbes, oxidants) and control scaling.
Pressure Vessel	The cylindrical housing (typically FRP or SS) that contains multiple membrane elements in series under high operating pressure.

Profiling	A troubleshooting technique involving measuring permeate conductivity from individual pressure vessels to locate leaks (O-rings, damaged elements).
Probing	An advanced troubleshooting technique using a special probe to collect permeate samples from specific points within a pressure vessel's permeate tube to pinpoint leaking elements.
Purified Water (PW)	A grade of pharmaceutical water defined by USP/EP, often produced using RO followed by polishing (e.g., EDI or IX).
Recovery Rate (%)	The percentage of feed water converted into permeate (Recovery = (Qp / Qf) * 100). Limited by scaling potential in the concentrate.
Rejection (Salt Rejection, %)	The percentage of dissolved solutes in the feed water that are removed by the membrane (Rejection = [1 - (Cp / Cf_avg)] * 100). Indicates purification efficiency.
Remineralization	Post-treatment step for RO permeate used for drinking water, involving adding back calcium and magnesium minerals to improve taste and reduce corrosivity.
Scaling	Fouling caused by the precipitation of inorganic mineral salts (e.g., $CaCO_3$, $CaSO_4$, SiO_2) on the membrane surface when their solubility limits are exceeded in the concentrate.
Seawater (SW)	Water from the ocean, typically with high TDS (>30,000 mg/L). Requires high-pressure RO (SWRO) for desalination.
Sedimentation (Clarification)	A pretreatment process allowing heavier flocculated particles to settle out by gravity in a large basin.
Selectivity	The membrane's ability to allow water passage while blocking solutes (salts). Also refers to potential future ability to separate specific ions.
Semi-Permeable Membrane	A membrane that allows certain molecules (like water) to pass through while blocking others (like dissolved salts). The basis of osmosis and RO.

Silica (SiO_2)	A dissolved compound (silicic acid) present in many waters that can polymerize and form difficult-to-remove scale in RO concentrate, often limiting recovery.
Silt Density Index (SDI)	An empirical measure of the fouling potential of feed water due to fine suspended solids and colloids. Target <3-5 for RO feed. Measured by plugging rate of a 0.45μm filter.
Skid (RO Skid)	The structural metal frame on which RO system components (pumps, vessels, filters, piping, controls) are typically mounted for factory assembly and installation.
Sodium Bisulfite (SBS)	A reducing agent ($NaHSO_3$) commonly injected into feed water to chemically remove residual chlorine/oxidants before TFC membranes.
Solution-Diffusion Model	The dominant theoretical model explaining transport across dense RO membranes, involving dissolution of water/solute into the polymer, diffusion across, and desorption.
Spiral Wound Module (SWM)	The most common RO element configuration, where membrane leaves and spacers are rolled around a central permeate tube. Offers high packing density.
Stage (in an Array)	A group of pressure vessels operating in parallel, where the concentrate from one stage feeds the next stage in series. Used to manage flow velocity and concentration.
Stiff & Davis Stability Index (S&DSI)	A calculated index, similar to LSI but adjusted for higher ionic strength, used to predict $CaCO_3$ scaling tendency in brackish water or concentrate.
Supersaturation	The condition where the concentration of dissolved ions exceeds the equilibrium solubility limit of a mineral salt, creating the driving force for scaling.

Thin-Film Composite (TFC) Membrane	The standard modern RO membrane construction, featuring an ultrathin, selective polyamide layer formed via interfacial polymerization on a microporous support (e.g., polysulfone). High performance but chlorine intolerant.
Thin-Film Nanocomposite (TFN) Membrane	An advanced TFC membrane incorporating nanoparticles (e.g., zeolites, CNTs, MOFs) into the polyamide layer to potentially enhance permeability, selectivity, or fouling resistance.
Total Dissolved Solids (TDS)	The total concentration of all dissolved substances in water (mg/L or ppm). Determines osmotic pressure and is a key measure of water salinity and permeate quality.
Total Organic Carbon (TOC)	A measure of the total amount of carbon in dissolved and suspended organic matter in water (mg/L). Used as an indicator of potential organic fouling or biofouling substrate.
Train (RO Train)	A complete, independent RO system (pretreatment, pumps, membranes, controls). Large plants use multiple parallel trains for capacity and redundancy.
Troubleshooting	The systematic process of diagnosing and resolving problems in an RO system based on data analysis, testing, and logical deduction.
Tubular Module	A membrane module configuration where the membrane is located inside porous tubes. Offers excellent fouling resistance but very low packing density and high cost. Used for very difficult feeds.
Turbidity	A measure of water cloudiness caused by suspended particles (NTU). Low turbidity (<0.1-0.2 NTU) is desired for RO feed, but SDI is a better predictor of fouling potential.
Ultrafiltration (UF)	Membrane process with pores ~0.01-0.1 μm, removing suspended solids, colloids, bacteria, viruses. Increasingly used as high-quality RO pretreatment, providing consistently low SDI water.

Ultrapure Water (UPW)	Water of extremely high purity (resistivity >18 Megohm-cm, very low TOC, ions, particles) required for semiconductor, pharmaceutical, and laboratory applications. Produced using multi-stage treatment including double-pass RO.
Variable Frequency Drive (VFD)	An electronic device used to control the speed of AC motors (like the RO feed pump), allowing efficient control of flow/pressure and soft starting.
Water For Injection (WFI)	The highest grade of pharmaceutical water, defined by USP/EP, requiring extremely low bacterial endotoxins. Traditionally made by distillation, increasingly by membrane methods post-RO.
Zero Liquid Discharge (ZLD)	A wastewater management strategy aiming to eliminate liquid effluent discharge by treating and recycling water and concentrating waste streams into solid residue. RO is a key component in ZLD systems.

Appendix C: Index

(Conceptual index referencing primary chapter discussions)

- Water Quality Parameters (Ch. 15)
- Water Potential (Ch. 2)
- Zero Liquid Discharge (ZLD) (Ch. 13, 17)

www.ingramcontent.com/pod-product-compliance
Lightning Source LLC
LaVergne TN
LVHW090314210926
848480LV00001B/239